Parameter Identification and
Monitoring of Mechanical Systems
under Nonlinear Vibration

Related titles:

Modelling and simulation of integrated systems in engineering
(ISBN 9780857090782)

Advanced engineering design
(ISBN 9780857090935)

Mechatronics and manufacturing engineering
(ISBN 978 0 85709 150 5)

Parameter Identification and Monitoring of Mechanical Systems under Nonlinear Vibration

Juan Carlos Jauregui

AMSTERDAM • BOSTON • CAMBRIDGE • HEIDELBERG • LONDON
NEW YORK • OXFORD • PARIS • SAN DIEGO
SAN FRANCISCO • SINGAPORE • SYDNEY • TOKYO
Woodhead Publishing is an imprint of Elsevier

WP
WOODHEAD
PUBLISHING

Woodhead Publishing is an imprint of Elsevier
80 High Street, Sawston, Cambridge, CB22 3HJ, UK
225 Wyman Street, Waltham, MA 02451, USA
Langford Lane, Kidlington, OX5 1GB, UK

Notice
No responsibility is assumed by the publisher for any injury and/or damage to
persons or property as a matter of products liability, negligence or otherwise,
or from any use or operation of any methods, products, instructions or ideas
contained in the material herein. Because of rapid advances in the medical
sciences, in particular, independent verification of diagnoses and drug dosages
should be made.

British Library Cataloguing-in-Publication Data
A catalogue record for this book is available from the British Library

Library of Congress Control Number: 2014948444

ISBN 978-1-78242-165-8 (print)
ISBN 978-1-78242-166-5 (online)

For information on all Woodhead Publishing publications
visit our website at http://store.elsevier.com/

Typeset by RefineCatch Limited, Bungay, Suffolk

Transferred to Digital Printing in 2014

To Elsa, Alonso, Hernan and to every engineer that works for building a better world

Table of contents

List of figures

About the author

Juan Carlos Jauregui was born in Mexico City on 12 May 1960. He obtained his BS degree in Mechanical Engineering from the National University of Mexico (UNAM) in 1983, and obtained his MS degree in 1984 from the same university. He studied his PhD at the University of Wisconsin-Milwaukee, where he graduated in 1986. His areas of expertise are machine design, structural analysis and mechanical vibrations.

Professor Jauregui works as a Professor at Universidad Autonoma de Queretaro where he conducts research in design and dynamics of machinery. He worked at CIATEQ (the Advanced Technology Center) from 1989 to 2011 and has been responsible for the design of a large number of automatic, tailor made machines which have been installed in different industries. Among these projects, he was in charge of the design and construction of a parallel robot for positioning the secondary mirror of the Large Milimetric Telescope (www.lmtgtm.org). He has been involved in the development of monitoring systems based on vibration analysis, and on this topic he has written several papers and the book *Mechanical Vibrations of Discontinuous Systems* (Nova Publishers). He is involved in the development of monitoring systems of machine tools. Up to now he has written more than 50 papers in international journals and congress.

Professor Jauregui belongs to several professional organizations such as ASME (American Society of Mechanical

Engineers), the Mexican Society of Mechanical Engineering, the Academy of Engineering (Mexico) and IFToMM. He is a member of the National Research System, the most prestigious research evaluation program in Mexico, and has won many prizes and recognitions.

Introduction

DOI: 10.1533/9781782421665.1

Abstract: Mechanical systems are complex in nature, they integrate several elements whose individual dynamics are generally nonlinear. The identification of these parameters requires a deep understanding of its dynamic behavior; therefore it is important to introduce basic concepts of dynamics. In this chapter, we will see the derivation of equations of motion from a classical dynamics point of view. Concepts such as generalized coordinates, constraints, the D'Alembert principle, Euler-Lagrange's and Hamilton's equations are described. These concepts will be used throughout the book and are the basis for understanding mechanical systems. In this chapter the concept of phase diagram, or phase plane, is fully developed since it is one of the techniques that identifies a nonlinear behavior and a stable behavior.

Key words: dynamics, Euler-Lagrange's equation, Hamilton's equation, phase diagram.

Evolution implies motion, and motion is the essence of any mechanical system. The way we understand, and as a consequence, design and improve mechanical systems is through the analysis of their motion. For doing this, we

1

simplify the analysis dividing the system into a set of simple elements interconnected, and then we can apply what we know about the motion of simple elements. The way those simple elements interconnect define the nature of every mechanical system.

The basic physics of particle and body motion grounds the theory for mechanical systems, but the challenge lays on the identification of the system's parameters. Although we can establish mathematical models based on the equations of motion, we have to be sure that these models represent the actual behavior of the mechanical system. Therefore we can say that the knowledge of the systems depends on the identification of its parameters and the relationship among its elements.

In order to understand the most advanced techniques for the identification of a system's parameters, it is important to know the background of body motion. In this chapter, the basic body motion theories are presented and I have selected the Lagrange's equation and the Hamilton's principle for developing the rest of the chapters.

Dynamics

Equation of motion

The motion of a particle, a system of particles or a body is determined by applying Newton's law of motions.[1] A mechanical system can be idealized as a set of N particles, where a particle is ideally a body with its mass concentrated at a point. Therefore, the motion of a particle can be found from the motion of a point. Since the point has no dimensions, its motion represents a unique curve in the space and it has no rotation associated with it.

The equation of motion of a system with N particles can be found by applying Newton's law to each individual particle.

The first concept to be introduced is the linear momentum: for a single particle with mass m

$$\bar{p} = m\bar{v} \tag{I.1}$$

where $\bar{v}$ is the velocity vector of a particle along the motion curve. It is determined from the inertial reference frame.

Figure I.1 describes the motion curve in a reference frame, where r is the vector describing the position of a particle at instant t and v is the velocity vector.

The inertial frame of reference is a basic concept in the Newtonian dynamics, and is assumed as the origin of the coordinate system with zero motion, and it is the "most convenient" origin of the coordinate system.

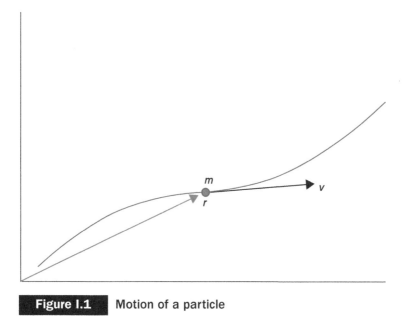

Figure I.1 Motion of a particle

Then, the position of the ith particle at a certain time t will be the vector $\bar{\rho}$, and its velocity is found as:

$$\bar{v} = \frac{d\bar{\rho}}{dt} = \dot{\bar{\rho}}$$ [I.2]

From Newton's law of motion, if a force F is applied to the ith particle, its linear momentum will change as:

$$\bar{F} = \frac{d(m\dot{v})}{dt}$$ [I.3]

The force acting on a body will be classified as:

- contact force
- body or field force.

Contact forces are applied by another body making direct contact and the body forces are acting without direct contact (such as gravity, the effect of electric fields, magnetic fields or other possible energy field).

If a Cartesian coordinate system is defined, equation I.3 can be represented as three nonlinear differential equations, which are a function of time, position and velocity. In many cases there is no close solution for these equations, and they have to be solved numerically. Finally, for a system on N particles there will be 3N nonlinear ordinary differential equations.

Generalized coordinates

Different sets of coordinates can be used to express the position of a point with respect to the inertial frame. Each set has a particular number of coordinates and a different number of constraints. But, for every set of coordinates, the number of coordinates minus the number of constraints must be equal to the degrees of freedom of the system. If two or more sets of coordinates are defined for a system, there must be a transform operator that transforms one set of coordinates into the other.

The parameters that describe the configuration of the mass points of the dynamic system are known as the generalized coordinates. They may include translations and rotations of different points of the system. The easiest way to define a generalized coordinate system is through the solution of those parameters that have geometrical significance, or if they are associated with a point motion. It is recommended that the generalized coordinates form an independent base. In this way, if the generalized coordinates define the configuration of the system without surpassing any constraint, then the generalized coordinate system is equal to the number of degrees of freedom.

The position of the N particle in a Cartesian coordinate system $(x_1, \ldots x_{3N})$ can be represented as a function of the generalized coordinates q_i and time as:

$$x_1 = f_1(q_1, \ldots, q_n, t)$$
$$x_2 = f_2(q_1, \ldots, q_n, t)$$
$$x_{3N} = f_{3N}(q_1, \ldots, q_n, t) \qquad [I.4]$$

where n is the number of generalized coordinates.

The number of constraints associated with the generalized coordinates is determined from the constraints of the system. If the number of constraints of the system is l and the number of constraints of the generalized coordinates is m, then the following relation must hold:

$$3N - l = n - m \qquad [I.5]$$

In order to have a unique correspondence between the generalized coordinates and the Cartesian system, the Jacobian determinant of the transformation must be non-zero:

$$\frac{\partial(f_{1,...,}f_{3N})}{\partial(q_{1,...,}q_{n})} \neq 0 \qquad\qquad\qquad [I.6]$$

Example:

Assume that a particle rotates around the inertial frame describing a circle (Figure I.2).

The position of a point will be:

$$x_1 = R\cos\theta$$
$$x_2 = R\sin\theta \qquad\qquad\qquad [I.7]$$

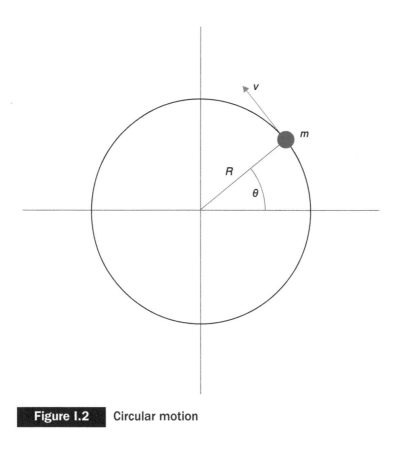

Figure I.2 Circular motion

Since the distance between the inertial frame and the particle is always the same, the constraint of the system is:

$$x_1^2 + x_2^2 = R^2 \qquad\qquad \text{[I.8]}$$

in order to transform the Cartesian coordinates into a generalized coordinate system defined as:

$$q_1 = \theta$$
$$q_2 = R \qquad\qquad \text{[I.9]}$$

Then the transformation function will be:

$$x_1 = f_1 = q_2\cos(q_1)$$
$$x_2 = f_2 = q_2\sin(q_1) \qquad\qquad \text{[I.10]}$$

The Jacobian determinant will be:

$$\frac{\partial f_1}{\partial q_1} = -q_2\sin(q_1)$$

$$\frac{\partial f_1}{\partial q_2} = \cos(q_1)$$

$$\frac{\partial f_2}{\partial q_1} = q_2\cos(q_1)$$

$$\frac{\partial f_2}{\partial q_2} = \sin(q_1) \qquad\qquad \text{[I.11]}$$

Thus,

$$\begin{vmatrix} -q_2\sin(q_1) & \cos(q_1) \\ q_2\cos(q_1) & \sin(q_1) \end{vmatrix} = -q_2 \qquad\qquad \text{[I.12]}$$

Constraints

There are two types of constraints and their definition depends on the generalized coordinate system. Physically a constraint is the function that describes the motion restriction of the system's particles. And for each motion restriction there is a restrictive force or constraint force.

If a system is defined by n generalized coordinates and it has m independent constraints, then there will be m constraints functions

$$\varphi_j(q_i, \ldots, q_n)(j=1, \ldots, m) \qquad [I.13]$$

A system will be defined as a hollonomic system if

$$\varphi_j(q_i, \ldots, q_n) = 0 \,\forall j \qquad [I.14]$$

Otherwise the system is considered as a non hollonomic system.

Example

If a disc rotates in a plane without slip following any trajectory, then the constraint will be of the form (Figure I.3):

$$ds = rd\varphi$$
$$dx = r\sin(\alpha)d\varphi$$
$$dy = r\cos(\alpha)d\varphi \qquad [I.15]$$

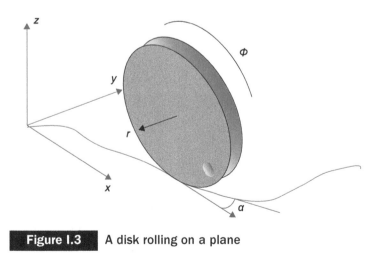

Figure I.3   A disk rolling on a plane

There are four coordinates (x,y,r,φ) and two constraints (Eq. 15), then, there is a set of differential equations of the form:

$$\sum_{i=1}^{n} a_{ij} dq_i + a_{jt} dt = 0 \qquad [\text{I.16}]$$

If equation I.16 is non integrable, then the constraints are non hollonomic.

Physically, in the example, the disc will follow any arbitrary path, and if it returns to any previous position, the contact point on the surface of the disc will be different. In other words, if we mark a point A in the disc surface and the disc follows an arbitrary path and returns to the original point x_0, y_0, point A will not be in contact with the plane.

D'Alembert principle

In order to apply the static equilibrium principle to a dynamic system, the inertial forces can be considered as external forces if:

$$\Sigma(\bar{F}_i + \bar{R}_i) - m_i\ddot{\bar{r}}_i = 0 \qquad [\text{I.17}]$$

where F_i are the external forces, R_i are the constraint forces and $m_i\ddot{\bar{r}}_i$ are inertial forces.

Generalized forces

The external forces are defined in the same coordinate frame as the system of particles. Therefore, if we define the motion of a particle in the generalized coordinate system, the same transformation must be applied to the external forces. It can be demonstrated that the generalized forces can be determined by:

$$Q_j = \sum_{i=1}^{3N} F_i \frac{\partial x_i}{\partial q_j} \qquad [\text{I.18}]$$

Energy and momentum

The potential energy is defined as the work done by a force that depends only on the initial and final position and it is independent of the velocity or time. Therefore, the forces associated to the potential energy are, in a Cartesian system:

$$F_x = -\frac{\partial V}{\partial x}$$

$$F_y = -\frac{\partial V}{\partial y}$$

$$F_z = -\frac{\partial V}{\partial z} \qquad [\text{I.19}]$$

where $V(x,y,z)$ is the potential energy function.

The work W done by a force $\bar{F}$ along an infinitesimal displacement dr is:

$$dV = \bar{F} \cdot d\bar{r} = F_x dx + F_y dy + F_z dz \qquad [\text{I.20}]$$

Substituting equation I.19:

$$dV = -\frac{\partial V}{\partial x} dx - \frac{\partial V}{\partial y} dy - \frac{\partial V}{\partial z} dz = -dV \qquad [\text{I.21}]$$

Integrating equation I.21 between positions A and B

$$V = -\int_A^B dV = V_A - V_B \qquad [\text{I.22}]$$

This means that the potential energy is independent of the specific path connecting points A and B. If the external forces are transformed into the generalized forces, then the potential energy can be written as:

$$V = \int_A^B \bar{Q} \cdot d\bar{q} \qquad [\text{I.23}]$$

The kinetic energy is determined from the inertial forces. The work done by the inertial forces (from Newton's law) along an infinitesimal displacement $d\bar{r}$ is:

$$dT = \frac{d\bar{p}}{dt} \cdot d\bar{r} \qquad [I.24]$$

If the mass of the particle is constant, the incremental work is:

$$dT = m\frac{d\bar{v}}{dt} \cdot d\bar{r} \qquad [I.25]$$

Expressing $\bar{v}$ in Cartesian coordinates

$$dT = m\left(\frac{dv_x}{dt}dx + \frac{dv_y}{dt}dy + \frac{dv_z}{dt}dz\right) \qquad [I.26]$$

Rearranging terms it is found that

$$dT = m\bar{v} \cdot d\bar{v} \qquad [I.27]$$

Integrating the work done by an inertial force between positions A and B

$$T = \int_A^B m\bar{v} \cdot d\bar{v} = \tfrac{1}{2}m(v_A^2 - v_B^2) \qquad [I.28]$$

In a system with N particles, the kinetic energy T will be equivalent to:

$$T = \tfrac{1}{2}\sum_{i=1}^{3N} m_i v_i^2 = \tfrac{1}{2}\sum_{i=1}^{3N} m_i \dot{x}_i^2 \qquad [I.29]$$

Expressing the velocity in terms of the generalized coordinate system

$$\dot{x}_i = \sum_{j=1}^{n} \frac{\partial x_i}{\partial q_j}\dot{q}_j + \frac{dx_i}{dt} \qquad [I.30]$$

Therefore, the kinetic energy of the system will be:

$$T = \frac{1}{2}\sum_{i=1}^{3N} m_i \left(\sum_{j=1}^{n} \frac{\partial x_i}{\partial q_j} \dot{q}_j + \frac{dx_i}{dt} \right)^2 \qquad \text{[I.31]}$$

Lagrange's equation

The Lagrange equation is derived by applying variational calculus and the first law of thermodynamics, or the conservation of energy. Since the energy of a particle, or a system of particles, is a function of its position, velocity and the time, it can be expressed as:

$$F = F(q_1, \ldots, q_n, \ddot{q}_1 \ldots, \ddot{q}_n, t) \qquad \text{[I.32]}$$

where q_i and $\ddot{q}_i$ are the generalized coordinates and their time derivatives.

According to the first law of thermodynamics, the system of N particles will move from position A to position B following the less energy trajectory; therefore, the equation of motion can be obtained from the stationary trajectory of the energy function.

The Langrange's function expresses the conservation of the mechanical energy as:

$$L = T - V \qquad \text{[I.33]}$$

Since the kinetic and potential energies are function of the generalized coordinates and their time derivatives, the mechanical energy of a particle moving from point A to point B can be expressed as:

$$L = \int_A^B F(\bar{q}, \dot{\bar{q}}, t) dt \qquad \text{[I.34]}$$

The first law of thermodynamics implies that L must be a stationary function, which any path variation must be zero at the points A and B, therefore:

12

$$\delta \int_{t_0}^{t_1} L\, dt = \delta I = 0 \qquad\qquad \text{[I.35]}$$

where, according to the calculus of variations

$$\delta I = \left(\frac{\partial I}{\partial \alpha}\right)_{\alpha=0} \delta\alpha = 0 \qquad\qquad \text{[I.36]}$$

Since L is zero at points A and B, they are independent of α, then the derivative of L must be always zero and the time interval between points A and B is t_0 to t_1, then:

$$\frac{dI}{d\alpha} = \int_{t_0}^{t_1} \left(\frac{\partial F}{\partial q_i}\frac{\partial q_i}{\partial \alpha} + \frac{\partial F}{\partial \dot{q}_i}\frac{\partial \dot{q}_i}{\partial \alpha}\right) dt = 0$$

$$\frac{dI}{d\alpha} = \int_{t_0}^{t_1} \left(\frac{\partial F}{\partial q_i}\frac{\partial q_i}{\partial \alpha}\right) dt + \int_{t_0}^{t_1} \left(\frac{\partial F}{\partial \dot{q}_i}\frac{\partial \dot{q}_i}{\partial \alpha}\right) dt = 0 \qquad \text{[I.37]}$$

Integrating the second term by parts

$$\int_{t_0}^{t_1} \left(\frac{\partial F}{\partial \dot{q}_i}\frac{\partial \dot{q}_i}{\partial \alpha}\right) dt = \left(\frac{\partial F}{\partial \dot{q}_i}\frac{\partial q_i}{\partial \alpha}\right)\bigg|_{t_0}^{t_1} - \int_{t_0}^{t_1} \frac{\partial q_i}{\partial \alpha}\frac{d}{dt}\left(\frac{\partial F}{\partial \dot{q}_i}\right) dt \qquad \text{[I.38]}$$

then

$$\frac{dI}{d\alpha} = \int_{t_0}^{t_1} \left(\frac{\partial F}{\partial q_i}\frac{\partial q_i}{\partial \alpha} - \frac{\partial q_i}{\partial \alpha}\frac{d}{dt}\left(\frac{\partial F}{\partial \dot{q}_i}\right)\right) dt + \left(\frac{\partial F}{\partial \dot{q}_i}\frac{\partial q_i}{\partial \alpha}\right)\bigg|_{t_0}^{t_1} = 0 \quad \text{[I.39]}$$

Since the second term is zero at the end points, the argument of the integral must be zero

$$\frac{\partial F}{\partial q_i} - \frac{d}{dt}\left(\frac{\partial F}{\partial \dot{q}_i}\right) = 0 \qquad\qquad \text{[I.40]}$$

and $F = L$, therefore the Lagrange's equation is:

$$\frac{d}{dt}\left(\frac{\partial L}{\partial \dot{q}_i}\right) - \frac{\partial L}{\partial q_i} = 0 \qquad\qquad \text{[I.41]}$$

The analysis and modeling of dynamic systems can be done from a Lagrangian approach or from a Hamiltonian approach. The Lagrangian approach describes how position and velocity change in time. The Hamiltonian approach describes how position and momentum change in time. The position and momentum of a particle specifies a point in a space called the "phase space", "phase plane", "phase diagram", among others.

A particle traces out a path in a space R^n

$$q_i : R \rightarrow R^n \qquad [I.42]$$

where R represents time domain, R^n represents the space domain and q_i represents the generalized coordinates of a particle at an instant t.

Hamilton's equation

Hamilton's principle is one of the most important variational principle of dynamics. It states that a hollonomic system follows a minimum energy trajectory. Expressing the energy of N particles system in terms of the momentum and the potential energy

$$H(\bar{q}, \bar{p}) = \frac{p^2}{2m} + V \qquad [I.43]$$

and

$$\frac{\partial H(q,p)}{\partial p_i} = \frac{p_i}{m} \qquad [I.44]$$

$$\frac{\partial H(q,p)}{\partial q_i} = \frac{\partial V(q)}{\partial q} \qquad [I.45]$$

Thus

$$\frac{d}{dt} q_i(t) = \frac{\partial H}{\partial p_i} (q(t), p(t)) \qquad\qquad \text{[I.46]}$$

$$\frac{d}{dt} p_i(t) = -\frac{\partial H}{\partial q_i} (q(t), p(t)) \qquad\qquad \text{[I.47]}$$

where the dyad $(q(t), p(t))$ represents the phase space of a particle, and $(q, p) \in \mathcal{R}^n x \mathcal{R}^n$.

If the phase space can be represented as a smooth function $\varphi \colon \mathcal{R}^n x \mathcal{R}^n \to \mathcal{R}$, then it represents the system's evolution in time. Thus, for a system with n particles

$$\frac{d\varphi}{dt} = \sum_i \frac{\partial\varphi}{\partial q_i} \frac{dq_i}{dt} + \frac{\partial\varphi}{\partial p_i} \frac{dp_i}{dt} \qquad\qquad \text{[I.48]}$$

Using Hamilton's equation

$$\frac{d\varphi}{dt} = \sum_i \frac{\partial\varphi}{\partial q_i} \frac{\partial H}{\partial p_i} - \frac{\partial\varphi}{\partial p_i} \frac{\partial H}{\partial q_i} \qquad\qquad \text{[I.49]}$$

The equation of motion of a particle following a simple harmonic is represented as

$$m\ddot{q}(t) + kq(t) = 0 \qquad\qquad \text{[I.50]}$$

with its well-known solution

$$q(t) = A sin(\omega t) + B cos(\omega t) \qquad\qquad \text{[I.51]}$$

$$p(t) = m(A\omega cos(\omega t) - B\omega sin(\omega t)) \qquad\qquad \text{[I.52]}$$

where

$$\omega = \sqrt{\frac{k}{m}} \qquad\qquad \text{[I.53]}$$

Applying initial conditions, at $t = 0$

$$B = q(0) \qquad\qquad \text{[I.54]}$$

$$A = \frac{p(0)}{m\omega} \qquad\qquad\qquad\qquad [I.55]$$

The Hamiltonian can be written as:

$$H(q,p) = \frac{1}{2}\left(\frac{p^2}{m} + kq^2\right) \qquad\qquad\qquad [I.56]$$

If we define the field vector operator as

$$\nu_H = p\frac{\partial}{\partial q} - q\frac{\partial}{\partial p} \qquad\qquad\qquad [I.57]$$

and the flow field is found as

$$\varphi_t = [q(t), p(t)] \qquad\qquad\qquad\qquad [I.58]$$

in this case

$$\varphi_t = \left[q(0)\cos(\omega t) + \frac{p(0)}{m\omega}\sin(\omega t), p(0)\cos(\omega t) - (q(0)m\omega)\sin(\omega t)\right] \quad [I.59]$$

This flow field represents an ellipse at any time t. (Figure I.4)

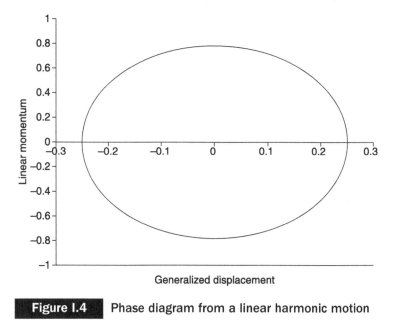

Figure I.4 Phase diagram from a linear harmonic motion

The dynamic stability is determined from Liouville's theorem (the phase space volume occupied by a collection of systems evolving according to Hamilton's equations of motion will be preserved in time):

$$\frac{dH}{dt} = \frac{\partial H}{\partial t} + \sum_{i=1}^{n} \left(\frac{\partial H}{\partial q_i} \dot{q}_i + \frac{\partial H}{\partial q_i} \dot{p}_i \right) = 0 \qquad [I.60]$$

It can be shown that

$$\sum_{i=1}^{n} \left(\frac{\partial^2 H}{\partial q_i \partial p_i} - \frac{\partial^2 H}{\partial q_i \partial p_i} \right) = 0 \qquad [I.61]$$

This conservation law states that a phase diagram volume will be preserved in time; this is the statement of Liouville's theorem.

In the following chapters, the use of the Hamilton's equation will be presented. This equation is a powerful tool for determining the stability of mechanical systems, and it is the basis for defining the phase diagram.

Note

1 Some of the concepts included in this chapter may be familiar to the reader, but they are included in order to define a notational framework for the entire book

Linear vibrations

DOI: 10.1533/9781782421665.19

Abstract: Once the fundamental concepts of dynamics are defined, it is important to describe mechanical vibrations. Mechanical vibrations cover a broad field of knowledge; thus, in this chapter we start defining the basic aspects of linear vibrations. The concepts are derived from the equation of motion, which is formulated from the energy methods introduced in the previous chapter. The equation of motion is a second order differential equation, and the linear term comes from the assumption that the coefficients of this equation are constant. The equation of motion depends on the generalized coordinates, and for each generalized coordinate we set a degree of freedom. The basic concepts are introduced with the analysis of a single degree of freedom system. For this system, three cases are presented: free undamped vibration, forced vibration and damped vibration. From the forced vibration concept, a definition of transmissibility is introduced and finally a system with multiple degrees of freedom is presented. The example included in this chapter is a gear box with four degrees of freedom; this example will be used thoughout the book.

Key words: free vibration, forced vibration, damping, transmissibility, multiple degrees of freedom.

1.1 Introduction

The dynamics of any machinery is perceived as noise or vibration, any machine vibrates and if we understand its vibrations, we can know its conditions. In general, engineering analysis is done assuming the behavior of a system is linear. Although it is not true, this approximation is good enough to solve any operating problem or to design a new machine. In this sense, the following section will describe the theory and applications of linear vibrations.

The concepts presented next are derived from previous definitions, and they will be centered around the general engineering concepts of the dynamic of elastic systems. It is assumed that a machine can be represented as a set of masses connected through elastic elements (springs) and dissipative elements (dampers). Since we are assuming that the behavior of such systems is linear, the movement of the masses will be represented as oscillations.

The development of vibrarion sensors allows the measurement and collection of vibration data – the most common sensors are piezoelectic accelerometers, capacitive accelerometers, velocity sensors and displacement sensors. Additionally, the development of signal analysis methods simplifies the diagnosis of machinery. The construction of frequency spectrums based on the Fast Fourier Transform (FFT) has become a well known procedure in industrial applications. Nowadays there are several companies that offer vibration analysis systems that can be implemented for the diagnosis of machinery. That is the reason for defining the math basis of vibration analysis.

1.2 Single degree of freedom

A simplified representation of a machine is a mass supported by a spring and a damper. This representation includes the main elements and helps understanding the dynamic behavior of complex systems. Figure 1.1 shows the representation of a mass m moving in the vertical direction, the movement is restricted by a spring k and the movement is damped by a viscous damper with a constant c.

The equation of motion is derived using previous concepts. In this case, the generalized coordintates will be the displacement x, and the potential and kinetic energies are:

$$V = \frac{1}{2}kx^2 \qquad\qquad [1.1]$$

$$T = \frac{1}{2}m\dot{x}^2 \qquad\qquad [1.2]$$

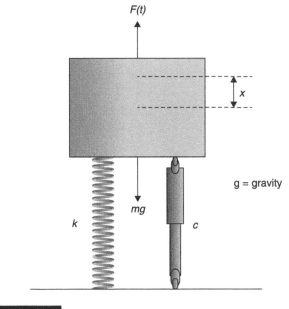

Figure 1.1   Simplified model of a machine

The damper dissipates energy, therefore the Lagrange equation of the system will be:

$$\frac{d}{dt}\left(\frac{\partial L}{\partial \dot{q}_i}\right) - \frac{\partial L}{\partial q_i} = Q \qquad [1.3]$$

where

$$L = T - V \qquad [1.4]$$

and

$$Q = f(t) - c\dot{x} \qquad [1.5]$$

The equation of motion is found substituting equations 1.4 and 1.5 into 1.3:

$$m\ddot{x} + c\dot{x} + kx = f(t) \qquad [1.6]$$

There are two solutions, if $f(t) = 0$ the system will freely vibrate and it is known as 'free vibration'. In this case the solution of the differential equation is

$$x = e^{pt}\left[x_0 \cos(qt) + \frac{\dot{x}_0}{q}\sin(qt)\right] \qquad [1.7]$$

where

$$p = -\frac{c}{2m} \qquad [1.8]$$

$$q = \sqrt{\left(\frac{k}{m}\right) - \left(\frac{c}{2m}\right)^2} = \omega_d \qquad [1.9]$$

and x_0 and $\dot{x}_0$ are the initial conditions.

If we define $\xi = \dfrac{c}{2m\omega_n}$ as the damping coefficient

$$\omega_d = \omega_n\sqrt{1-\xi^2} \qquad [1.10]$$

then the most simple motion occurs when $c = 0$ in which case

$$p = 0$$

and

$$q = \sqrt{\left(\frac{k}{m}\right)} = \omega_n$$

and the mass moves according to a simple harmonic motion

$$x = x_0 \cos(\omega_n t) + \frac{\dot{x}_0}{\omega_n} \sin(\omega_n t) \qquad [1.11]$$

Figure 1.2 represents the simple harmonic motion of a system when $\dot{x}_0 = 0$. This type of motion is known as 'Undamped Free Vibration'.

If $c \neq 0$ then the mass displacement damps until it rests. The form of this type of motion is represented in Figure 1.3. The function that envelopes the sinusoidal wave is known as the logarithm decrement, and it is possible to measure the damping coefficient of a system.

The amplitude of two consecutive peeks x_1 and x_2 are separated by a period T. Then, from equation 1.7

$$x_1 = e^{\frac{c}{2m} t_1} \qquad [1.12]$$

$$x_2 = e^{\frac{c}{2m}(t_1 + T)} \qquad [1.13]$$

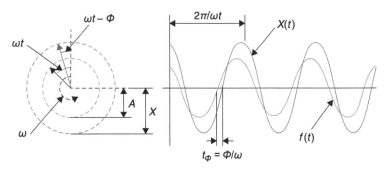

Figure 1.2 Harmonic motion of a single degree system and phase angle

Thus,

$$\ln\left(\frac{x_1}{x_2}\right) = \frac{c}{2m}T = \delta \qquad [1.14]$$

where

$$T = \frac{2\pi}{\omega_d} \qquad [1.15]$$

and the damping is

$$c = 2\omega_n m \frac{\delta}{\sqrt{2\pi + \delta^2}} \qquad [1.16]$$

Or in a parametric form:

$$\xi = \frac{\delta}{\sqrt{2\pi + \delta^2}} \qquad [1.17]$$

In general, $f(t)$ is assumed to be sinusoidal. In this case, the solution of the differential equation (Eq. 1.6) will be

$$m\ddot{x} + c\dot{x} + kx = f_0\cos(\Omega t) \qquad [1.18]$$

The solution of equation 1.18 is

$$x = \frac{f_0\cos(\Omega t - \phi)}{\sqrt{\left(\frac{k}{m} - \Omega^2\right)^2 + \left(\frac{c}{m}\right)^2}} \qquad [1.19]$$

Or

$$\frac{x}{f_0/m} = \frac{1}{\sqrt{\left(1 - \left(\frac{\Omega}{\omega_n}\right)^2\right)^2 + \left(\frac{2\xi\Omega}{\omega_n}\right)^2}}\cos(\Omega t - \phi) \qquad [1.20]$$

where

$$tg(\phi) = \frac{2\xi\left(\dfrac{\Omega}{\omega_n}\right)}{1 - \left(\dfrac{\Omega}{\omega_n}\right)^2} \qquad [1.21]$$

From equation 1.20 the transfer function of the dysnamic system is definded as

$$H(\Omega) = \frac{1}{\sqrt{\left(1 - \left(\dfrac{\Omega}{\omega_n}\right)^2\right)^2 + \left(\dfrac{2\xi\Omega}{\omega_n}\right)^2}} \qquad [1.22]$$

Figure 1.3 shows the amplitude of the transfer function, and Figure 1.4 describes the variations of the phase angle.

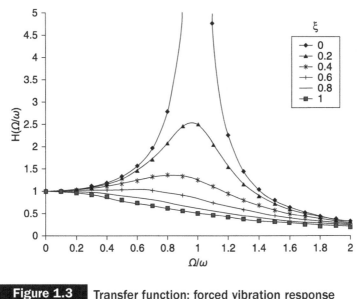

 **Figure 1.3** Transfer function: forced vibration response

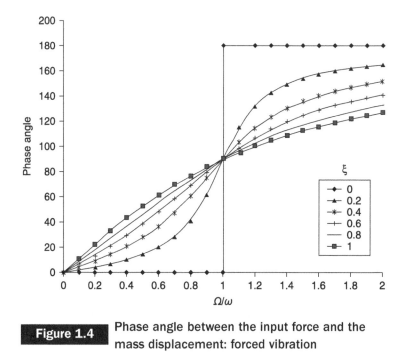

Figure 1.4   Phase angle between the input force and the mass displacement: forced vibration

The critical speed is the natural frequency of the system. Beyond this value, the phase angle between the excitation force and the mass displacement is closed to 180°. If the excitation force comes from the foundation, then the movement of the mass is the result of the trasmissibility of the supports, in this case we represent the supports as a linear spring and a linear damper. The force transmitted by these elements is

$$f_k = kx_b \qquad\qquad [1.23]$$

$$f_c = c\dot{x}_b \qquad\qquad [1.24]$$

Then, the equation of motion becomes

$$m\ddot{x} + c\dot{x} + kx = kx_b + c\dot{x}_b \qquad\qquad [1.25]$$

Assuming that the base has an armonic motion, then $x_b = A\cos(\Omega t)$ and $\dot{x}_b = -A\Omega\sin(\Omega t)$. Thus, the equation of motion is

$$m\ddot{x} + c\dot{x} + kx = A(k\cos(\Omega t) - c\Omega\sin(\Omega t)) \qquad [1.26]$$

or

$$\ddot{x} + 2\xi\omega_n\dot{x} + \omega_n x = A(\omega_n\cos(\Omega t) - 2\xi\omega_n\Omega\sin(\Omega t)) \qquad [1.27]$$

The solution of the equation of motion will be $x = X\cos \times (\Omega t - \phi)$.

The relation between the excitation amplitude and the mass displacement amplitude is the trasmissibility of the foundation, and it is

$$|T| = H(\Omega) = \frac{\sqrt{1 - \left(\frac{2\xi\Omega}{\omega_n}\right)^2}}{\sqrt{\left(1 - \left(\frac{\Omega}{\omega_n}\right)^2\right)^2 + \left(\frac{2\xi\Omega}{\omega_n}\right)^2}} \qquad [1.28]$$

This equation is the basis for a foundation design, and the design criterion will be to have the minimum transmitted force. Thus, it can be shown that the magnitude of the transmitted force $|T| > 1$ will occur if $\frac{\Omega}{\omega_n} > \sqrt{2}$. That means that the excitation frequency must be greater than the natural frequency. But in many cases this type of design is not possible and the operating frequencies are a lot lower than the natural frequencies. Thus, the stiffness and damping characteristics should be modified by adding a soft material between the vibration source and the object that will be isolated. Since there are two design parameters, the problem is underdetermined; therefore, a range of frequencies and an amplitude threshold must be defined. Figure 1.5 shows a design chart for a support. The desired amplitude is the input

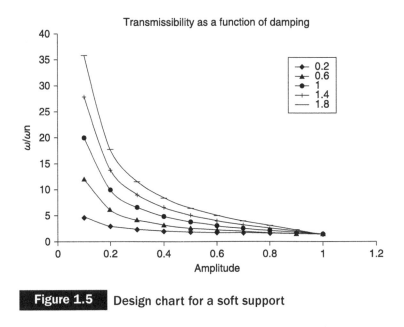

Figure 1.5 Design chart for a soft support

data, and the output is the range of frequencies that the isolation will absorb.

1.3 Multiple degrees of freedom

Although many engineering problems can be solved with a single degree of freedom model, complex systems have to be analyzed as a multibody system. The system is divided into a set of point inertias (masses or moments of inertia) connected through springs and dampers. Each inertia represents a generalized coordinate and the equation of motion is derived using Lagrange's equation. The concept is introduced with an example. Figure 1.6 shows a gear transmission, where the gear teeth and the rolling bearings are represented by linear springs. In the following chapters they will be modeled as nonlinear elements.

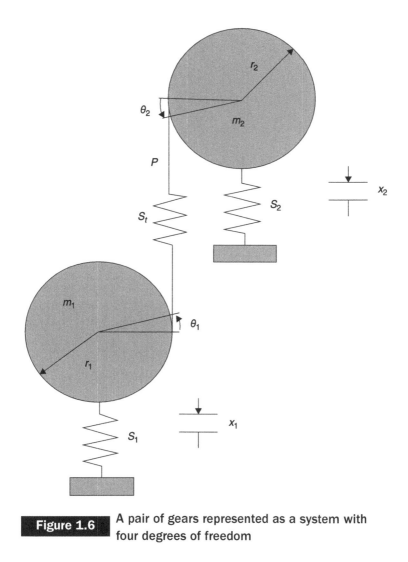

Figure 1.6 A pair of gears represented as a system with four degrees of freedom

Representing the gear box as a lumped-mass system, four generalized coordinates are identified: $\{x_1, x_2, \theta_1, \theta_2\}$. The equation of motion is obtained using Lagrange's equation and the dynamic behavior of the pair of gears consists of four differential equations:

$$\ddot{x}_1 + \frac{k_1}{m_1}x_1 + \frac{k_e}{m_1}[(x_1 - x_2) + (r_1\theta_1 + r_2\theta_2)] = \frac{P_1(t)}{m_1} \qquad [1.29]$$

$$\ddot{x}_2 + \frac{k_2}{m_2}x_2 + \frac{k_e}{m_2}[(x_1 - x_2) + (r_1\theta_1 + r_2\theta_2)] = \frac{P_2(t)}{m_2} \qquad [1.30]$$

$$\ddot{\theta}_1 + \frac{k_e r_1}{J_1}[(x_1 - x_2) + (r_1\theta_1 + r_2\theta_2)] = \frac{r_1 P_e(t)}{J_1} \qquad [1.31]$$

$$\ddot{\theta}_2 - \frac{k_e r_2}{J_2}[(x_1 - x_2) + (r_1\theta_1 + r_2\theta_2)] = \frac{-r_2 P_e(t)}{J_2} \qquad [1.32]$$

The system of equations can be represented in a matrix form as:

$$\begin{bmatrix} 1 & \cdots & 0 \\ \vdots & \ddots & \vdots \\ 0 & \cdots & 1 \end{bmatrix} \begin{Bmatrix} \ddot{x}_1 \\ \ddot{x}_2 \\ \ddot{\theta}_1 \\ \ddot{\theta}_2 \end{Bmatrix} + \begin{bmatrix} k_{11} & \cdots & k_{14} \\ \vdots & \ddots & \vdots \\ k_{41} & \cdots & k_{44} \end{bmatrix} \begin{Bmatrix} x_1 \\ x_2 \\ \theta_1 \\ \theta_2 \end{Bmatrix} = \begin{Bmatrix} P_1 \\ P_2 \\ P_e \\ -P_2 \end{Bmatrix} \qquad [1.33]$$

or in a simplified form

$$[I]\{\ddot{\bar{X}}\} + [K]\{\bar{X}\} = \{\bar{p}\} \qquad [1.34]$$

The homogenous solution of equation (1.34) is found assuming that $\{\bar{p}\} = 0$ and $x_i = X_i\cos(\omega_i t - \phi_i)$. Thus, the matrix equation will be:

$$[-\omega^2[I] + [K]]\{X\} = 0 \qquad [1.35]$$

The solution is found solving the eignevalues problem, and the roots of the polynomial correspond to the natural frequencies of each generalized coordinates

$$\det|[K] - \omega^2[I]| = 0 \qquad [1.36]$$

Solving the eigenvalue problem, a fourth order polynomial is found:

$$\lambda^4 + \lambda^3 \left(k_e \left(\frac{r_2^2}{J_2} - \frac{r_1^2}{J_1} \right) - k_e \left(\frac{1}{m_1} + \frac{1}{m_2} \right) - k_1 \left(\frac{1}{m_1} \right) - k_2 \left(\frac{1}{m_2} \right) \right)$$

$$+ \lambda^2 \left(k_1 k_2 \left(\frac{1}{m_1 m_2} \right) + k_1 k_e \left(\frac{1}{m_1 m_2} + \frac{r_1^2}{m_1 J_1} - \frac{r_2^2}{m_1 J_2} \right) \right.$$

$$\left. + k_2 k_e \left(\frac{-1}{m_1 m_2} + \frac{r_1^2}{m_2 J_1} - \frac{r_2^2}{m_2 J_2} \right) \right)$$

$$+ \lambda \left(k_1 k_2 k_e \left(\frac{r_2^2}{m_1 m_2 J_2} - \frac{r_1^2}{m_1 m_2 J_1} \right) \right) \tag{1.37}$$

where $\lambda = \omega^2$.

The particular solution is found numerically using the Runge-Kutta method. The external forces are the excitation forces caused by the rotation of the rolling bearings and the gear teeth action, they are approximated as:

$$P_e(t) = (F_p + F_r)\cos(\Omega t) + (f_p + F_\varphi)\cos(N\Omega t) \tag{1.38}$$

where F_r is the gear run out, F_p is the accumulated pitch error, F_φ is the profile error, and f_p is the pitch error.

$$P_j(t) = F_{ir}\cos(\Omega_{ir}) + F_{or}\cos(\Omega_{or}) + F_c\cos(\Omega_c)$$
$$+ F_{re}\cos(\Omega_{re}) \tag{1.39}$$

where $P_j(t)$ are the bearing excitation forces ($j = 1,2$).

The four frequencies are calculated with the following equations.

Contact frequency between the roller element and the internal track Ω_{ir}:

$$\Omega_{ir} = \frac{N}{2} \left[1 + \frac{db}{Dp}\cos(\alpha_b) \right] \Omega \tag{1.40}$$

Contact frequency between the roller element and the external track Ω_{or}:

$$\Omega_{or} = \frac{N}{2}\left[1 - \frac{db}{Dp}\cos(\alpha_b)\right]\Omega \qquad [1.41]$$

The casing has a frequency Ω_c:

$$\Omega_c = \frac{1}{2}\left[1 - \frac{db}{Dp}\cos(\alpha_b)\right]\Omega \qquad [1.42]$$

and the roller spin frequency is Ω_{re}:

$$\Omega_{re} = \frac{Dp}{db}\left[1 - \left(\frac{db}{Dp}\cos(\alpha_b)\right)^2\right]\Omega \qquad [1.43]$$

where db is the roller diameter and Dp is the pitch diameter, α_b is the axial contact angle.

As an example, a gear box is simulated to previous concepts. In the first part, the natural frequencies are found using the eigenvalue solution of equation 1.37. Then, the forced vibration solution of the multiple degree of freedom problem is solved using the Runge-Kutta algorithm.

The gear data listed in the table below and with these values, the natural frequencies are $\omega_1 = 937.0$ rad/s, $\omega_2 = 662.5$ rad/s, $\omega_3 = 371.9$ rad/s.

	Gear 1	Gear 2
Mass (m)	4.87 kg	10.57 kg
Moment of inertia (J)	0.0011 kg-m^2	0.004652 kg-m^2
Base radius (r)	0.031853 m	0.0483285 m
Bearing stiffness (K)	3.66×10^6 N-m	4.1410^6 N-m
Gear mesh stiffness (Ke)		4.37×10^5

The solution for the forced vibration part is calculated assuming that all the excitation forces are acting on the gear box. In this way, the characteristic spectrum of an industrial gear box is illustrated, and the particular frequencies are identified.

Nonlinear vibrations

DOI: 10.1533/9781782421665.35

Abstract: Modern machinery combine higher operating speeds with lighter elements, and this combination is one of the reasons why nonlinear vibrations occur frequently. The scope of equations of motion increases tremendously since many mechanical elements show different types of nonlinearities; even more, the mathematical model representing the equation of motion can include several orders and an infinitive number of possible coefficients. Therefore, in this chapter the most typical models are included: a nonlinear pendulum, the Van der Pol and Duffings oscillators, self-excited vibrations and friction. These models describe most of the elements that contribute to nonlinearities in mechanical systems, such as gears, bearings and friction elements. Solutions are found using numerical methods such as the Runge-Kutta method.

Key words: nonlinear oscillators, Van der Pol, Duffing, self-excited vibrations, friction.

2.1 Introduction

Most of the dynamic systems can be studied using linear models. Nevertheless, modern machinery operates at higher velocities and tends to be as light as possible. The combination of these two factors makes it difficult to simulate modern machinery with a linear model. As an example, automotive gear transmissions are designed as light as possible, and with several gear reductions. In order to reduce the weight, new transmissions are designed with hollow shafts. This concept modifies the natural frequencies and the inclusion of several gear reductions increases the number of external loads. Thus, there is a higher probability of having excitation frequencies near natural frequencies.

The most common typical nonlinear elements in a mechanical system are gears, rolling bearings, belts and friction. In the case of gears, the stiffness of the transmission depends upon the relative position of the teeth in contact. Rolling bearings have a kinematic behavior that modifies the dynamics of the system and the reduce friction based on the movement of the roller around a track. This movement is a combination of rotation of the roller around its geometric center and an epicyclical translation. The load carried by the rollers changes as a function of the epicyclical movement, therefore the stiffness of the system varies as a function of the angular speed. In the case of belts, the stiffness of the system depends upon the material properties, which in general are non-constant. Friction is a complex phenomenon that modifies the damping (dissipation) function of the system.

In general, the dynamic response of a nonlinear system can be represented as a differential equation of the form:

$$\ddot{x} + f(\dot{x}, x, t) = 0 \qquad\qquad [2.1]$$

In this case, the superposition principle is no longer valid, and the system shows harmonic distortion. Harmonic distortion is one of the clearest indicators of nonlinearity. It is a straightforward consequence of the principle of superposition. If the excitation force is sinusoidal the displacement must be sinusoidal with the same frequency. Distortion is easily identified because the displacement functions are harmonic but not sinusoidal. Thus, one way of identifying a slender beam with nonlinear behavior is through the application of sinusoidal excitation and comparing the 'shape' of the displacement: if it is not sinusoidal, then the system is nonlinear. In practice it is not that simple. Other signal analysis techniques, such as wavelets, can identify this 'shape' variation.

Another technique to test if a system behaves nonlinearly is through reviewing the homogeneity and distortion of the transfer function in the frequency domain. Homogeneity means that the transfer function will be invariant under changes of the level of excitation. This is not always true, since there are some nonlinear systems which show homogeneity. The reason is that this property is a weaker condition than superposition. Only if a control system is employed to maintain a constant excitation force spectrum can nonlinearity be detected through homogeneity.

The other important property, which can be used to detect nonlinearities, is reciprocity. When employing reciprocity all parameters must be the same, e.g. the output must be acceleration and the input must be forces. Thus, if it is a linear system, the stiffness matrix will be symmetric.

Analytical procedures are difficult, with few exact solutions, thus numerical solutions (such as Runge-Kutta) are the most appropriate way to analyze nonlinear mechanical systems.

In a conservative system, the total energy remains constant. Expressing the system's total energy as the sum of kinetic and potential energy:

$$\frac{1}{2}m\dot{x}^2 + V(x) = E \qquad [2.2]$$

where E remains constant at any time.

If we define the state variable $y = \dot{x}$ then

$$y = \sqrt{2(E - V(x))/m} \qquad [2.3]$$

Regardless of the shape of $V(x)$, the trajectories of the phase diagram of the system will be symmetric with respect to x. Thus the differential equation of motion will have the form of

$$\ddot{x} = f(x) \qquad [2.4]$$

or

$$\dot{x} d\dot{x} = f(x) dx \qquad [2.5]$$

Integrating equation 2.5:

$$\frac{\dot{x}^2}{2} - \int_0^x f(x)\,dx = \frac{E}{m} \qquad [2.6]$$

Then, it can be proved that

$$\frac{V(x)}{m} = -\int_0^x f(x)\,dx$$

Thus

$$f(x) = -\frac{1}{m}\frac{dV(x)}{dx} \qquad [2.7]$$

and the phase diagram becomes

$$\frac{dy}{dx} = \frac{f(x)}{y} \qquad [2.8]$$

As an example, consider that the potential energy has the following function:

$$V(x) = \alpha x + \beta x^2 + \gamma x^3 \qquad [2.9]$$

The value of E depends upon the initial conditions $x(0)$ and $y(0)$. At any position E must be greater than $V(x)$, otherwise y will be imaginary. Therefore,

$$E = \frac{1}{2}m[\dot{x}(0)]^2 + V(x(0))$$
[2.10]

As shown in Figure 2.1, for $E > 0.72$ there is no equilibrium in the system. For $E = 0.5$, there are two close loops around $x = -4.1$ and $x = 1.0.$, and for E between 0.5 and 0.72 there is only one loop for each value. The period associated with each loop can be found integrating equation 2.3:

$$\tau = \int_{x_1}^{x_2} \frac{dx}{\sqrt{2(E - V(x))/m}}$$
[2.11]

where x_1 and x_2 are the intersections of the phase diagram with the x axis.

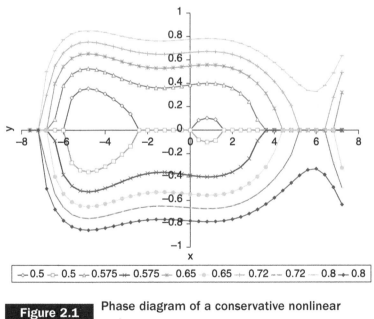

Figure 2.1 Phase diagram of a conservative nonlinear system

This analysis can be expressed in a general form. The singular points can be found expressing equation 2.8 as

$$\frac{dy}{dx} = \frac{F(x,y)}{G(x,y)} \qquad [2.12]$$

If $F(x,y)$ and $G(x,y)$ are expanded in therms of a Taylor's series around the singularity points

$$F(x,y) = F(x_s,y_s) + \left(\frac{\partial F}{\partial x}\right)_s x + \left(\frac{\partial F}{\partial y}\right)_s y + h.o.t$$

$$G(x,y) = G(x_s,y_s) + \left(\frac{\partial G}{\partial x}\right)_s x + \left(\frac{\partial G}{\partial y}\right)_s y + h.o.t \qquad [2.13]$$

Since $\dfrac{\partial F}{\partial x'}, \dfrac{\partial F}{\partial y}, \dfrac{\partial G}{\partial x}$ and $\dfrac{\partial G}{\partial y}$ are constant at the singularity points, equation 2.12 can be expressed as:

$$\frac{dy}{dx} = \frac{ax + by}{cx + dy} \qquad [2.14]$$

Equation 2.14 can be rearranged as:

$$\frac{dx}{dt} = ax + by$$

$$\frac{dy}{dt} = cx + dy \qquad [2.15]$$

The solution of equation 2.15 is found solving the eigenvalue problem:

$$\begin{vmatrix} (a-\lambda) & b \\ c & (d-\lambda) \end{vmatrix} = 0 \qquad [2.16]$$

The stability of the system depends on the values of a, b, c and d; thus:

If $(ae - bc) > \left(\dfrac{a+e}{2}\right)^2$ the motion is oscillatory.

If $(ae-bc) < \left(\dfrac{a+e}{2}\right)^2$ the motion is aperiodic.

If $(a+e) > 0$, the system is unstable.

If $(a+e) < 0$, the system is stable.

Another way to evaluate if the system is unstable is through the analysis of the number of loops, and their location in the phase diagram. This analysis will be shown in the following chapters.

An illustrative example is the oscillation of a pendulum (Figure 2.2) In this case, if the amplitude of the oscillation is large enough, then the equation of motion is:

$$\ddot{\theta} + \frac{g}{l}\sin(\theta) = 0 \tag{2.17}$$

Expressing the equation of motion in terms of equation 2.6:

$$\frac{\dot{\theta}^2}{2} = \frac{g}{l}\cos(\theta) + C \tag{2.18}$$

Depending on the value of C, the phase diagram will have the form shown in Figure 2.3. From this figure, it is clear that there are several attracting values, around 0, 2π, 4π, etc. (This redundancy results from the total rotation of the pendulum). For $C<1$, the solution is stable and always crosses the horizontal axis, at $C=1$, the pendulum can oscillate downwards or upwards depending on the initial conditions. For $C>1$ the solution never crosses the horizontal axis and there is no solution. It is important to emphasize that C depends upon the initial conditions.

One of the well-known nonlinear models is the Van der Pol equation. This equation represents those systems that, for large oscillations, the damping is positive, whereas for small oscillations the damping is negative. This phenomenon can be found in electric circuits containing

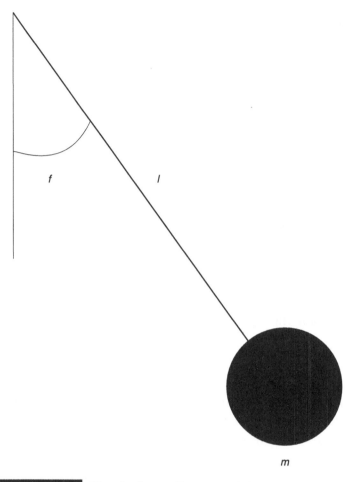

Figure 2.2 Sketch of a nonlinear pendulum

vacuum tubes. The differential equation that describes this type of systems is:

$$\ddot{x} - \zeta(1 - x^2)\dot{x} + \omega^2 x = 0 \qquad [2.19]$$

Thus, the equation of motion can be expressed as:

$$\frac{d\dot{x}}{dx} = \zeta(1 - x^2) - \frac{\omega^2 x}{\dot{x}} \qquad [2.20]$$

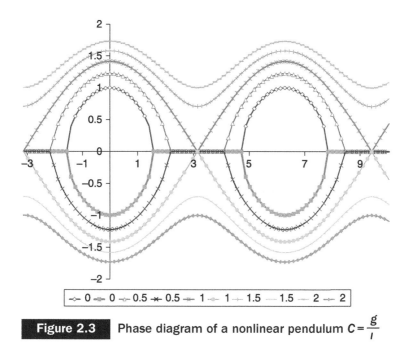

| --◇-- 0 | --▣-- 0 | --○-- 0.5 | --✳-- 0.5 | --◁-- 1 | --▷-- 1 | --+-- 1.5 | ----- 1.5 | ----- 2 | --△-- 2 |

Figure 2.3 Phase diagram of a nonlinear pendulum $C = \dfrac{g}{l}$

Equation 2.20 is a first order differential equation; it has a simple geometric interpretation, which will lead to a useful method for obtaining a practical solution without an explicit solution. If we define equation 2.20 as

$$y' = \frac{d\dot{x}}{dx} = H(x,y) \qquad\qquad [2.21]$$

and, if we assume that the function $H(x,y)$ is defined in a region of the phase diagram, then $H(x,y)$ can be plotted as curves in the phase diagram, assuming that it is constant. These constant curves are defined as isoclines. Therefore,

$$y' = \frac{d\dot{x}}{dx} = H(x,y) = k \qquad\qquad [2.22]$$

43

This yields to

$$\dot{x} = \frac{\omega^2 x}{\zeta(1-y^2)-k}$$

[2.23]

There are three possible solutions, if $\zeta=0$, then the equation represents a simple harmonic system, but if $\zeta>1$ shows a negative damping for small oscillations or a positive damping for large oscillations, than, the limit cycle will depend upon the initial conditions. Figures 2.4 and 2.5 show the phase diagram with different initial conditions, and it is clear that the solution depends on the initial conditions and it confirms the system's nonlinearities. The time response is represented in Figure 2.6; although the function is periodic, the response is non-harmonic.

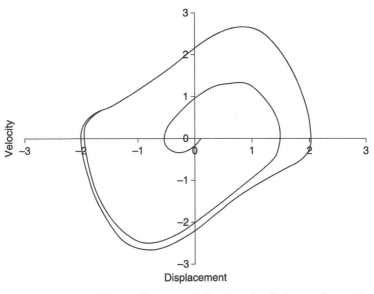

Figure 2.4 Phase diagram of the Van der Pol equation with initial conditions within the limit cycle

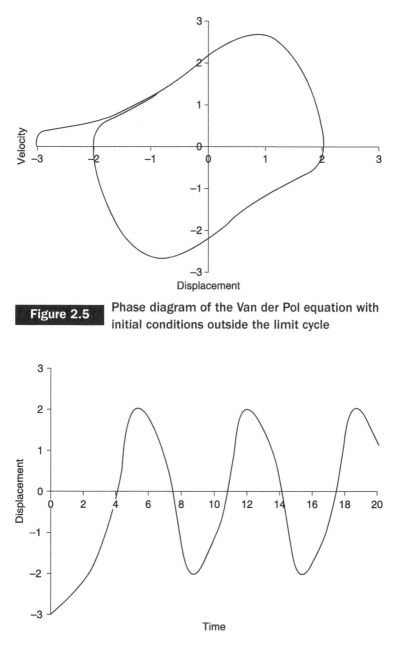

Figure 2.5 Phase diagram of the Van der Pol equation with initial conditions outside the limit cycle

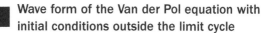

Figure 2.6 Wave form of the Van der Pol equation with initial conditions outside the limit cycle

2.2 Jump phenomena

One of the most common representations of a nonlinear system is the Duffing's equation. This equation considers that the system's stiffness varies as a function of the displacement. The mathematical representation is

$$\ddot{x} + \frac{c}{m}\dot{x} + \omega_0^2 x + \beta x^3 = A\cos(\omega t) \qquad [2.24]$$

If $\beta > 0$ then the system has a hardening stiffness and for $\beta < 0$ the system has a softening stiffness.

For illustration, it is considered that the damping coefficient is zero and there is no forced excitation. Then the equation can be expressed as:

$$\ddot{x} + \omega_0^2 x + \beta x^3 = 0 \qquad [2.25]$$

One of the procedures for solving this equation is with the perturbation method. This method is valid for small values of β. If we consider the initial conditions $x(0) = X$ and $\dot{x}(0) = 0$, then solution can be expressed as a polynomial series of the form

$$x(t) = x_0(t) + \beta x_1(t) + \beta^2 x_2(t) + \ldots + \beta^n x_n(t) \qquad [2.26]$$

and we know that the frequency of the system will vary with respect to the initial conditions and the parameter β

$$\omega^2 = \omega_0^2 + \beta \alpha_1 + \beta^2 \alpha_2 + \ldots + \beta^n \alpha_n \qquad [2.27]$$

Since β is small we can neglect the higher order terms

$$x(t) = x_0(t) + \beta x_1(t) \qquad [2.28]$$

and

$$\omega_0^2 = \omega^2 - \beta \alpha_1 \qquad [2.29]$$

Substituting this equation into the equation of motion:

$$\ddot{x}_0 + \beta \ddot{x}_1 + (\omega^2 - \beta \alpha_1)(x_0 + \beta x_1) + \beta(x_0 + \beta x_1)^3 = 0 \qquad [2.30]$$

Expanding terms:

$$\ddot{x}_0 + \omega^2 x_0 + \beta x_0^3 + 3\beta^2 x_0^2 x_1 + 3\beta^3 x_0 x_1^2 + \beta^4 x_1^3$$
$$- \beta \alpha_1 x_0 + \beta \ddot{x}_1 + \beta \omega^2 x_1 - \beta^2 \alpha_1 x_1 = 0 \qquad [2.31]$$

Since β is nonzero and small, then we separate equation 2.31 into

$$\ddot{x}_0 + \omega^2 x_0 = 0$$
$$\ddot{x}_1 + \omega^2 x_1 = \alpha_1 x_0 - x_0^3 \qquad [2.32]$$

The solution of the first term is define by imposing the initial conditions

$$x_0 = X \cos(\omega t) \qquad [2.33]$$

The second part of the equation will become

$$\ddot{x}_1 + \omega^2 x_1 = \alpha_1 X \cos(\omega t) - X^3 \cos^3(\omega t) \qquad [2.34]$$

Using the trigonometric identity

$$\cos^3(\omega t) = \frac{1}{4}(3\cos(\omega t) + \cos(3\omega t)) \qquad [2.35]$$

Then:

$$\ddot{x}_1 + \omega^2 x_1 = \left(\alpha_1 X - \frac{3}{4}X^3\right)\cos(\omega t) + \frac{1}{4}X^3 \cos(3\omega t) \qquad [2.36]$$

In order to have a non-resonant solution (the excitation frequency is equal to ω):

$$\alpha_1 X - \frac{3}{4}X^3 = 0 \qquad [2.37]$$

Therefore

$$\alpha_1 = \frac{3}{4}X^2 \qquad [2.38]$$

Thus, the frequency of the system will be:

$$\omega^2 = \omega_0^2 + \frac{3}{4}\beta X^2 \qquad\qquad [2.39]$$

Finally, the solution of the system is

$$x = X\left(1 - \frac{\beta X^2}{32\omega^2}\right)\cos(\omega t) + \frac{\beta X^3}{32\omega^2}\cos(3\omega t) \qquad [2.40]$$

If we consider a system with an external excitation, the dynamic equation will be:

$$\ddot{x} + \omega_0^2 x + \beta x^3 = F\cos(\omega t) \qquad\qquad [2.41]$$

Following a similar procedure, it can be shown that the solution is possible if:

$$\omega^2 = \omega_0^2 + \frac{3}{4}\beta X^2 - \frac{F}{X} \qquad\qquad [2.42]$$

Equation 2.42 can be rearrange as

$$\frac{3}{4}\beta\frac{X^3}{\omega_0^2} = \left(1 - \frac{\omega^2}{\omega_0^2}\right)X - \frac{F}{\omega_0^2} \qquad [2.43]$$

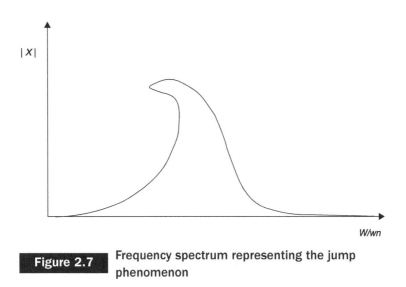

$|X|$

W/wn

Figure 2.7 Frequency spectrum representing the jump phenomenon

If we plot $|X|$ against ω/ω_0 we can see that there is a region where there are two possible solutions if $\beta < 0$.

2.3 Self-excited vibrations

There are mechanical systems in which the oscillations depend on the motion itself. One of the most representative examples is the formation of chatters in a machining process. The motion is induced by an excitation that is a function of the velocity and the displacement. If the motion increases the energy of the system, then it becomes unstable. Dry friction is another representative example of self-excited vibrations. It is an important source of mechanical damping in many physical systems. The viscous-like damping property suggest that many mechanical designs can be improved by configuring frictional interfaces in ways that allow normal forces to vary with displacement. Dry friction has two dominant components, one is associated to the displacement (static friction) and the other is associated to the velocity (kinetic friction).

Distinctions between coefficients of static and kinetic friction have been mentioned in the friction literature for centuries. Euler developed a mechanical model to explain the origins of frictional resistance. He arrived at the conclusion that friction during sliding motion should be smaller. The experiment proposed by Euler involved the sliding of a body down an inclined plane at slopes close to the critical slope at which sliding initiates. This, of course, would mean that, as soon as sliding initiates, a drop of friction force occurs, the difference between static and kinetic friction forces being responsible for the acceleration of the body down the inclined plane. This concept is also defined as stick-slip in sliding systems.

The distinction between static and kinetic friction was also a major topic of Coulomb's detailed experimental study. Coulomb's work is the first major reference dealing with the increase of the coefficient of static friction with increasing times of repose. He observed a dependence of the kinetic friction on the sliding velocity and a dependence of the static friction on the time of repose. However, for dry metal-to-metal interfaces all those distinctions or variations were absent or negligible.

In general, the coefficient of kinetic friction would be small and increasing with sliding velocity at low velocities. Then, at some velocity, it would achieve a maximum value after which it would decrease with the increase of speed. The sliding process is not a continuous one; the motion proceeds by jerks. The metallic surfaces 'stick' together until, as a result of the gradually increasing pull, there is a sudden break with a consequent very rapid 'slip'. The surfaces stick again and the process is repeated indefinitely. When the surfaces are of the same metal, the behavior is somewhat different. Large fluctuations in the friction still occur but they are comparatively slow and very irregular. The average value of the frictional force is considerably higher than that found for dissimilar metals and a well-marked and characteristic track is formed during the sliding.

It has also been observed that the frequency of the stick-slip motion increases with the increase of the driving velocity and that the maximum value of this frequency approaches the undamped natural frequency of the system, although in some cases the oscillation stops at a level well below that of the natural frequency. The most common model is due to Martins (Anderson and Ferri 1990; Oden and Martins 1985). His model considers a two-degree-of-freedom system, where the normal force between the sliding block is its weight, and it is free to separate from the sliding surface upwards.

The dynamic behavior of a single-degree-of-freedom system with amplitude and rate dependent friction forces is presented. A system with amplitude-dependent friction is more likely to experience intermittent sticking. If the system sticks a significant amount of time, the energy dissipation capability may be degraded. Hence, special care is taken in this analysis to examine sticking conditions (in the case of gear teeth action sticking occurs only for very high contact stresses). In general sticking can occur only when the sliding velocity is zero.

The extended friction law is:

$$F_\mu = \mu(C_0 + C_1|x| + C_2|\dot{x}|)sgn(\dot{x}) \tag{2.44}$$

where x represents the sliding displacement, $\dot{x}$ represents the sliding velocity, C_0 is the normal force, C_1 is the friction interface amplitude, C_2 is the friction interface velocity and μ is the coefficient of friction (in general it is equivalent to the static coefficient of friction).

The dynamic model for a single-degree-of-freedom system is represented as:

$$m\ddot{x} + c\dot{x} + kx + F_\mu = F_e\cos(\omega t) \tag{2.45}$$

The system is positively damped at all times and it is clearly stable in the sense of Lyapunov. However, the system is not asymptotically stable for $C_0 \neq 0$. This condition is identified from the phase diagram when $F_e = 0$ and the initial $x = 0.1$. The friction force F_μ has a particular behavior as shown in Figure 2.8.

The solution of the dynamic equation 2.45 is a typical solution for a self-excited vibration; it has a stable solution, but its behavior is non-harmonic. The best representation of this behavior is the phase diagram as shown in Figure 2.9. It can be seen that independently of the initial condition, the solution stays at the limit cycle, but it shape differs from an ellipse.

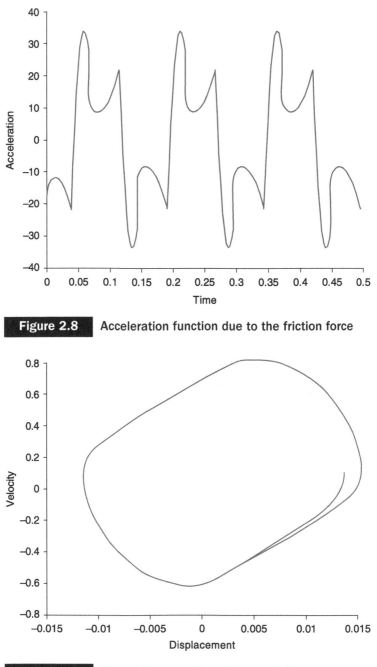

Figure 2.8   Acceleration function due to the friction force

Figure 2.9   Phase diagram of a self-excited vibration system

2.4 References

Anderson, J., and Ferri, A. (1990) Behavior of a single-degree-of-freedom-system with a generalized friction law, *Journal of Sound and Vibration*, Vol. 140(2), pp. 287–304.

Oden, J., and Martins, J. (1985) Models and computational methods for dynamic friction phenomena, *Computer Methods in Applied Mechanics and Engineering*, Vol. 52, pp. 527–634.

3

Signal processing

DOI: 10.1533/9781782421665.55

Abstract: Acquiring or generating vibration data is insufficient for analyzing a particular situation. Data by themselves will not determine sources and effects of a particular vibration case. Therefore, signal processing plays an important role in vibration analysis. Traditionally, Fourier Transform is applied to every situation and, for linear vibrations, it is sufficient to know the frequencies associated with vibration sources or the natural frequencies. But for nonlinear vibrations, Fourier Transform is not able to fully understand the phenomenon; therefore, in this chapter other signal analyses are introduced. Techniques such as Short Time Fourier Transform and the Continuous Wavelet Transform are described, applications of the phase diagram are included, and statistical techniques like the approximate entropy are presented. In many cases, more than one technique is needed in order to analyze a mechanical system.

Key words: Fourier Transform, Short Time Fourier Transform, Continuous Wavelet Transform, phase diagram, approximate entropy.

3.1 Introduction

Signal analysis is the basis for determining monitoring conditions and vibration effects. The most commonly measured variable is vibration. A vibration signal contains information regarding the dynamic behavior of a machine or structure. But this information is difficult to understand just from the vibration data, therefore, it has to be processed. Vibration signals are recorded from different types of sensors: piezoelectric accelerometers, capacitive accelerometers, strain gages, proximity sensors, and inductance velocity sensors, among others. The selection of the sensor is critical, it depends on several aspects, but there are two main parameters to be considered: the range of frequencies and the amplitude of the vibration. For very low frequencies, it is recommended to use displacement or velocity sensors. For higher frequencies, accelerometers are recommended. Regarding the amplitude, it is important to consider the gain of the sensors because it determines the quality of the 'shape' of the signal.

Since the behavior of a machine is time dependent, the monitoring signal is also time dependent to the dynamic condition. A detail analysis of the signal reveals the nature of the machine and helps to understand its condition. In order to overcome the difficulty of extracting the information from the vibration signal, the data must be transformed to a different domain. Most of the procedures to analyze this type of signal are based on the convolution theorem.

Convolution is a linear transformation of two periodic functions and it can be viewed as a cross correlation between them. The result of a convolution is the overlapped area between the two functions along the time. In other words, it determines the similarities between the two functions and, mathematically, it reflects the 'similarity' between the two functions.

The mathematical expression for the convolution of two continuous functions f and g is defined as:

$$(f * g)(t) = \int_{-\infty}^{\infty} f(\tau)g(t - \tau)d\tau \qquad [3.1]$$

where τ is the shifting term. Analyzing this expression, the convolution transform can be described as a weighted average of $f(\tau)$, where $g(t - \tau)$ is the weighting function, and it emphasizes the differences and similitudes between f and g.

The convolution is a linear transformation that satisfies the following algebraic properties:

- Commutativity, if f and g are two continuous functions, then:

$$f * g = g * f \qquad [3.2]$$

- Associativity, if f, g and h are three continuous functions, then:

$$f * (g * h) = (f * g) * h \qquad [3.3]$$

- Distributivity, if f, g and h are three continuous functions, then:

$$f * (g + h) = (f * g) + (f * h) \qquad [3.4]$$

- Scalar multiplication, if f and g are two continuous functions and α is a scalar, then:

$$\alpha(f * g) = (\alpha f) * g \qquad [3.5]$$

- Identity. Formally there is not an identity for every function, but for most of the periodic functions the identity function is the Dirac's delta distribution. Other functions admit an approximation to the identity.

$$f * \delta = f \qquad [3.6]$$

where

$$\int_{-\infty}^{\infty} \delta(\tau)d\tau = 1 \qquad\qquad [3.7]$$

- Inverse element, some function will admit an inverse element:

$$f * f^{-1} = \delta \qquad\qquad [3.8]$$

3.2 Convolution theorem

One way to simplify the application of the convolution transformation is through the application of the convolution theorem. Its demonstration goes beyond the scope of this chapter, but its application is very useful in the analysis of vibration signals.

If f and g are two continuous functions, then they can be transformed using the Fourier Transform (it will be defined in the next section). Then the convolution theorem states that

$$\langle f * g \rangle = \alpha F(\omega)G(\omega) \qquad\qquad [3.9]$$

where α is a normalizing scalar, $F(\omega)$ and $G(\omega)$ are the Fourier transforms of f and g.

With this theorem, the application of the convolution is simplified since there is no need for an explicit integration and it can be done in the frequency domain. Mathematically, it can be done by representing the time domain data as a series of well-known functions.

$$x(t) = \sum_{i=1}^{\infty} c_i f_i(t) \qquad\qquad [3.10]$$

The coefficients of the function $c_i f_i(t)$ are found comparing the original signal with a set of template functions. In this case, the coefficients of the function are calculated as:

$$c_i \equiv \langle x | \psi_i \rangle = \int_{-\infty}^{\infty} x(t)\psi_i^*(t)dt \qquad\qquad [3.11]$$

where $\psi_i^*(t)$ is the complex conjugated function of $\psi_i(t)$. In other words, c_i is the inner product of the data vector $x(t)$ and $\psi_i(t)$ and it determines a set of functions that are similar to $x(t)$. When $x(t)$ is similar to $\psi_i(t)$, the inner product is very large and the opposite holds true. Observing equation 3.2, there are an infinitive number of possible functions that will describe the original signal. Each set of functions is associated to different signal processing techniques such as the Fourier Transform, the Short Time Fourier Transform or the Wavelet Transform.

3.3 Fourier Transform

It is the most widely used signal analysis tool. In mechanical vibrations, it allows the identification of all the vibration sources exciting a machine. In many cases these sources of vibration have a harmonic behavior.

The Fourier Transform converts a data series $x(t)$ into a series of functions in the frequency domain. It is based on the Laplace Transform and on the concept that a periodic signal can be substituted by an infinite series of weighted sine and cosine functions. Using the Euler's complex number, the transformation function can be expressed as:

$$\psi_i^* = e^{-i\omega t} \qquad [3.12]$$

and the Fourier Transform is defined as

$$X(\omega) \equiv \left\langle x \middle| e^{-i\omega t} \right\rangle = \int_{-\infty}^{\infty} x(t) e^{-i\omega t} \, dt \qquad [3.13]$$

Modern measuring systems transform the transducer signal into a digital data vector. Thus, the time domain is discrete and has a time interval Δt equal to the sampling time (according to the data acquisition system). In this case, the

Fourier Transform will become a Discrete Fourier Transform which is defined as:

$$X_\omega = \frac{1}{N}\sum_{k=1}^{N-1} x_k e^{-i\omega k \Delta t}$$ [3.14]

Equation 3.14 represents the convolution between the vibration data and the sine and cosine functions. In this way, it is possible to decompose the original signal into a series of frequencies (Figure 3.2). These frequencies represent the dynamic behavior of the system, and (according to what was described in previous chapters) they are related to the natural frequency of the system and the applied excitation frequencies. In this way, the analysis of the original signal is reduced to the identification of the sources of vibration.

In a single degree of freedom with an external excitation as:

$$m\ddot{x} + c\dot{x} + kx = f_0\cos(\Omega t)$$ [3.15]

with initial conditions $x(0) = x_0$ and $\dot{x}(0) = 0$, the Fourier Transform will display two frequencies ω and Ω, where ω is determined from the system parameters as:

$$\omega^2 = \frac{k/m}{\frac{c^2}{4km - c^2} + 1}$$ [3.16]

and Ω is the excitation frequency. In the following example, where the system parameters are

$m = 10$

$k = 60000$

$c = 100$

$\Omega = 2\pi(50)$

then the natural frequency of the system will be $\omega = 77.3$ rad/s or 12.3 Hz. The original signal is plotted in Figure 3.2.

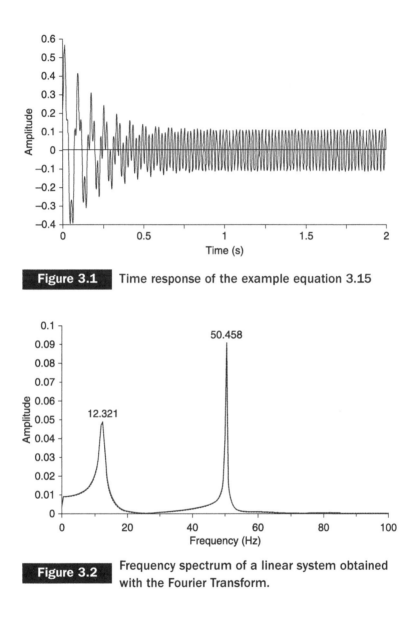

Figure 3.1 Time response of the example equation 3.15

Figure 3.2 Frequency spectrum of a linear system obtained with the Fourier Transform.

3.3.1 Aliasing and leakage

Aliasing refers to an effect that causes different signals to become indistinguishable (or aliases of one another). It also refers to the distortion of a signal when it is reconstructed from a sample data. In order to avoid aliasing, it was demonstrated that the sample frequency must be at least twice the maximum frequency of the vibration signal. Any frequency above this value will be eliminated from the spectrum.

The drawback of the Fourier Transform is that it eliminates the transient response and it is unable to deploy the

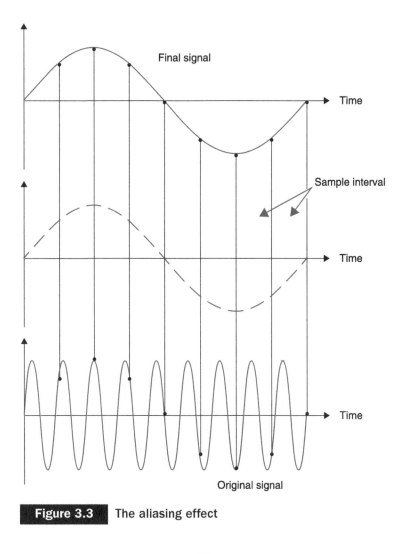

Figure 3.3 The aliasing effect

convolution of nonlinear responses. Even more, the Fourier Transform is unable to reveal how the signal frequency content varies with respect to time. In many systems, frequency time variations reveal its actual behavior, and reveal the nature of a nonlinear system. Due to this limitation, other signal analysis techniques have been developed.

3.4 Short Time Fourier Transform (STFT)

To overcome the limitations of the Fourier Transform, it is necessary to convert the signal data from a one dimension solution into a multiple dimension solution. The first method that solved this problem was presented by Dennis Gabor in 1946. He introduced a 'window' of certain length that slides along the signal to perform a time-localized Fourier transform. This is the concept of the Short Time Fourier Transform (STFT).

With this method a data signal can be segmented into a set of discrete Fourier transforms and, modifying the size of the window functions, the signal is decomposed into a 2D time-frequency analysis. In this way, the STFT has two arguments, time and frequency. In a time frequency map (also known as a 'spectrogram'), the time dependent components are separated from the steady state components, and both the transient components and the nonlinear effects can be identified as part of the entire map. The Short Time Fourier Transformation is obtained with

$$X(\omega, \tau) \equiv \left\langle x \middle| g(t) e^{-i\omega t} \right\rangle = \int_{-\infty}^{\infty} x(t) g(t - \tau) e^{-i\omega t} dt \qquad [3.17]$$

This equation can be interpreted as a measure of similarity between $x(t)$ and the modulated window function $g(t)$. This

63

segmentation produces a non-periodic function, which creates discontinuities at the boundaries; thus, the Fourier Transform creates large coefficients at high frequencies. To avoid these effects, the windowing concept is introduced. Instead of a rectangular segmentation, a smooth window function is employed to segment the original signal. The window function is almost one at the origin and zero at the edges. There are several functions reported in the literature and each one produces different results. The most commonly used functions are listed in the following table:

	$g(t)$
Gaussian	e^{-18t^2}
Hamming	$0.54+0.46\cos(2\pi t)$
Kaiser-Bessel	$0.402+0.498\cos(2\pi t)+0.099\cos(4\pi t)+0.001\cos(6\pi t)$

Each function has a different application. The Gaussian window identifies transient responses, Hamming allows the identification of narrowband random signals while the Kaiser-Bessel window is better for separating signals with very close frequencies, but with different amplitudes. As shown next, each function produces different time–frequency analyses. The STFT has limited resolutions in the time and frequency domains; thus the time and frequency resolutions cannot be chosen arbitrarily. To obtain finer time resolutions at higher frequencies will require keeping the same resolution at lower frequencies, which is redundant. The size of the window must keep a dimension in such a way that the lower size must be (Cohen 1989; Gao and Yan 2011):

$$\Delta\tau\Delta f \geq \frac{1}{4\pi} \qquad [3.18]$$

For example, the Gaussian function $g(t)=e^{-\alpha^2 t^2}$, where τ determines the size of the window.

Another limitation of the STFT is that the edges of each window produce artificial spectrum peaks. To avoid this, the windows are overlapped, and the overlap factor produces different analysis results. If the window function has a constant overlap property, the sum of all the individual Discrete Time–Function Transform will be the Discrete Time–Function Transform of the entire signal. The overlap concept is shown in Figure 3.4.

To illustrate the application of the STFT, equation 3.15 was analyzed with different window functions. Figure 3.5 shows a time–frequency map built with the Hamming function. This figure represents a 2D contour plot of the variations of the original signal as a function of time and frequency; the colors refer to the amplitude of vibration. In this figure, the contours are widely spread along the frequency axis, and only the transient response is easily identified (purple color). The steady state response (determined from the analytical solution) covers a wide range of frequencies and apparently they vary with time. We know that this is not true, and it can be noticed that the STFT is introducing an artificial variation in the analysis.

Figure 3.6 shows the same transformation using a Gaussian function with a 50% overlap. The transient response has a single gradient compared to the Hamming analysis. The steady state response also shows time variations due to the analysis. Figure 3.7 is the result of a similar analysis but without overlapping. In this contour map, it is clear that the overlapping reduces the time variations, and provides a smoother solution

Figure 3.8 shows the contour plot of a time–frequency map using a Gaussian function with a higher frequency resolution. In this case the low frequency response is lost, and the steady state response has a better resolution. Figures 3.9 and 3.10 show the analysis using a Kaiser-Bessel function. The best representation of the original signal is obtained with a

No overlap

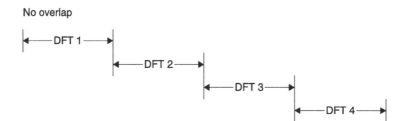

R/4 overlap

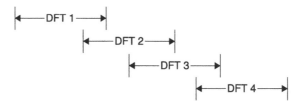

R/2 overlap

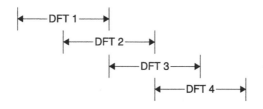

The parameter L

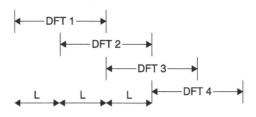

L is the number of samples between adjacent blocks.

Figure 3.4 The overlap concept in the STFT

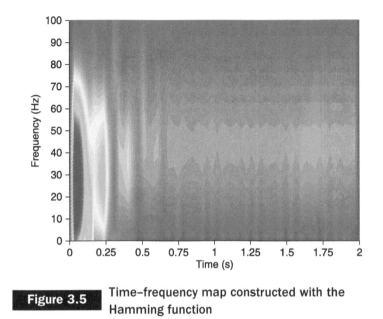

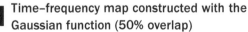

Figure 3.5 Time–frequency map constructed with the Hamming function

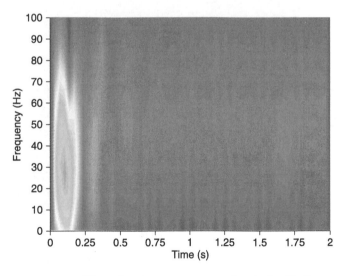

Figure 3.6 Time–frequency map constructed with the Gaussian function (50% overlap)

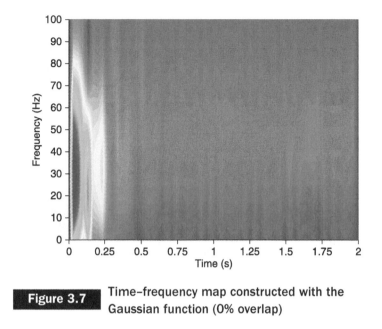

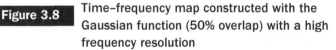

Figure 3.7 Time–frequency map constructed with the Gaussian function (0% overlap)

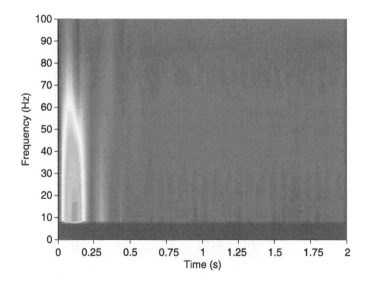

Figure 3.8 Time–frequency map constructed with the Gaussian function (50% overlap) with a high frequency resolution

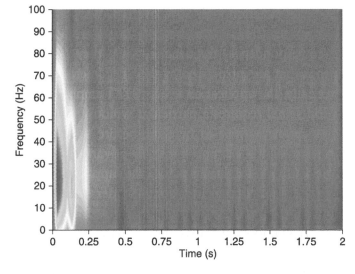

Figure 3.9 Time–frequency map produced with the Kaiser-Bessel function (0% ovelap)

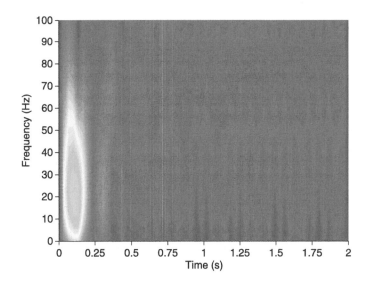

Figure 3.10 Time–frequency map constructed with the Kaiser-Bessel function (50% overlap)

Kaiser-Bessel function with 50% overlap. With this analysis both responses (transient and steady state) are identified at their correspondence frequencies, but they show time variations that are introduced by the STFT method.

Practically, the application of the STFT implies that each vibration signal requires a particular window function, and different analyses need to be done before setting the best parameters. If we do not know in advance the type of information we are looking for, the application of this method will require many iterations in order to set the proper window function. Low frequencies can hardly be identified with short windows, whereas short pulses can only poorly be localized in time with long windows.

The drawbacks of the STFT are overcome with the use of newer signal analysis techniques; the most appropriate method for the analysis of vibration signals is the Wavelet Transform.

3.5 Wavelet Transform

The word wavelet comes from the French term 'ondelette' which means small wave. Several researchers studied alternatives to the Fourier analysis, but Jean Morlet was the first to implement a technique for scaling and shifting the analysis window function. He analyzed acoustic echoes produced by acoustic impulses sent to the ground to determine the existence of oil beneath the Earth's crust. His experiments marked the beginning of wavelets, named by him and Alex Grossmann in the 1980s. Numerous researchers followed their work: Stromberg (1983) did some early work on discrete wavelet transform; Meyer (1993) and Malat (1998) worked on multi-resolution analysis; Newman (Gondek et al. 1994) worked on harmonic wavelet for vibration analysis; and Ingrid Daubechies (1992) proposed a family of wavelets based on the concept of orthogonal multi-resolution. She defined the concept of wavelet as '... a tool that cuts up data, functions or operators into different frequency components, and then studies each component with a resolution matched to its scale.' This concept has been implemented using simple digital filtering techniques, and they have been applied to many different fields, such as imaging processing, data condensation, solution of differential equations, etc. In the analysis of nonlinear dynamics, the Wavelet Transform will decompose the signal into its frequency components, preserving the time histories, and will discriminate those frequencies that remain constant along the entire response from those which vary along the time basis. The constant frequencies are related to a linear behavior, whereas the varying frequencies are related to the nonlinear terms.

The mathematical form of the Wavelet Transform is obtained from the convolution theorem:

$$X(s,\tau) \equiv \langle x | \psi(s,\tau) \rangle = \frac{1}{\sqrt{s}} \int_{-\infty}^{\infty} x(t)\psi^* \left(\frac{t-\tau}{s} \right) dt \qquad [3.19]$$

where $s > 0$ is the scaling parameter, is inversely proportional to the frequency, and determines the time and frequency scaling of the base function (namely mother wavelet). ψ^* is the complex conjugated function of the mother wavelet. For example, the Morlet mother wavelet is defined as

$$\psi \left(\frac{t-\tau}{s} \right) = e^{i 2\pi f_0 \left(\frac{t-\tau}{s} \right)} e^{-\alpha \frac{(t-\tau)^2}{s^2 \beta^2}} \qquad [3.20]$$

where f_0, α, and β are constant.

Equation 3.20 is formed by a sinusoidal function and a logarithm decrement function. The time interval will be calculated as:

$$\Delta t = \frac{s\beta}{2\sqrt{\alpha}} \qquad\qquad [3.21]$$

and the frequency interval as

$$\Delta f = \frac{\sqrt{\alpha}}{2\pi s\beta} \qquad\qquad [3.22]$$

Through variations of the scale and the time shift τ of the mother wavelet, the Wavelet Transform is able to decompose a signal into a time series along the entire spectrum. Small-scale values decompose high frequency components, whereas large-scale values decompose low frequency components

The basic properties of the Wavelet Transform are the translation and the dilatation process. The translation shifts the time axis by τ, and the dilation stretches or squeezes the wavelet by $1/s$. The application of these properties transforms the original signal into a set of scaled wavelets.

The factor $1/\sqrt{s}$ is included (eq. 3.19) in order to assure that the energy (ε) of the wavelet function is the same at any scale s.

$$\int_{-\infty}^{\infty}|\psi(t)|^2 dt = \frac{1}{s}\int_{-\infty}^{\infty}\left|\psi\left(\frac{t}{s}\right)\right|^2 dt = \varepsilon \qquad\qquad [3.23]$$

Any wavelet is a square integrable function that satisfies the admissibility conditions:

$$\int_{-\infty}^{\infty}\frac{|\psi(\omega)|^2}{\omega}d\omega < \infty \qquad\qquad [3.24]$$

where ψ is the Fourier Transform of the wavelet function. The admissibility condition implies that the Fourier Transform of the mother wavelet function is zero at zero frequency. The interpretation of this condition is that the wavelet has a band-pass filter like spectrum. The other implication is that the average value of $\psi(t)$ is zero:

$$\int_{-\infty}^{\infty} \psi(t)\,dt = 0 \qquad\qquad [3.25]$$

Thus, the wavelet must be oscillatory.

As long as the wavelet satisfies the admissibility condition, the inverse transform exists. This means that the original signal can be reconstructed from the wavelet transformation. The equation of the inverse transform is

$$x(t) = \left(\frac{1}{C_\psi} \int_{-\infty}^{\infty} \frac{1}{s^2}\,ds\right)\left(\int_{-\infty}^{\infty} X(s,\tau)\frac{1}{\sqrt{s}}\psi\left(\frac{t-\tau}{s}\right)d\tau\right) \qquad [3.26]$$

where $C_\psi = \int_{-\infty}^{\infty} \dfrac{|\psi(\omega)|^2}{\omega}\,d\omega$.

Some of the most common mother function are represented next.

3.5.1 Morlet function

This function represents the general behavior of a damped system and is expressed as:

$$\psi\left(\frac{t-\tau}{s}\right) = e^{i2\pi f_0\left(\frac{t-\tau}{s}\right)} e^{-\alpha\frac{(t-\tau)^2}{s^2\beta^2}} \qquad\qquad [3.27]$$

The graphical representation is shown in Figure 3.11.

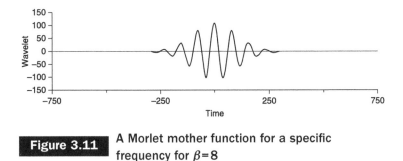

Figure 3.11 A Morlet mother function for a specific frequency for $\beta=8$

3.5.2 Gaussian function

The Gaussian function is determined from the cumulative probabilistic equation:

$$f(t) = e^{-it} \, e^{-t^2} \qquad [3.28]$$

and the mother wavelet is calculated as:

$$\psi(t) = C_N \frac{d^N f(t)}{dt^N} \qquad [3.29]$$

where N is a positive integer and C_N is a constant value that assures that $|\psi(t)|^2 = 1$. Figure 3.12 shows an example with $N = 2$.

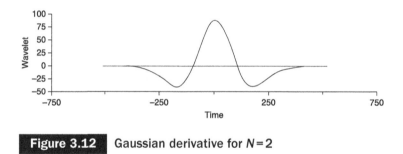

Figure 3.12   Gaussian derivative for $N = 2$

3.5.3 Paul's function

This mother function is determined from:

$$\psi(t) = \frac{2^N N! (1 - jt)^{-(N+1)}}{2\pi \sqrt{\frac{(2N!)}{2}}} \qquad [3.30]$$

Figure 3.13 shows the graphical representation of Paul's function for $N = 10$.

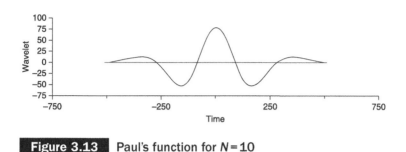

Figure 3.13 Paul's function for $N = 10$

3.5.4 Computer procedure

There are different alternatives for calculating the continuous wavelet transform. The easiest to implement is by applying the convolution theorem. The procedure is as follows:

1. Obtain the Fourier Transform of the original signal. Computationally it is obtained by applying the FFT $X(\omega) = F(x(t))$.

2. Set the scale factor to a minimum value s_0 and the time $t = \tau$.

3. Obtain the Fourier Transform of the mother wavelet $\psi(s, w) = F(\Psi(t))$.

4. Do inner product of the two transformations (convolution theorem).

5. Obtain the inverse Fourier Transform $X(s, t) = F^{-1}\{X(s, \omega)\}$.

6. Increment s and restart at step 2 until the desired number of scale factors is reached.

The other common alternative is to do the numerical integration of the convolution function in the time domain instead of doing the Fourier Transform. This procedure relays on the integration method.

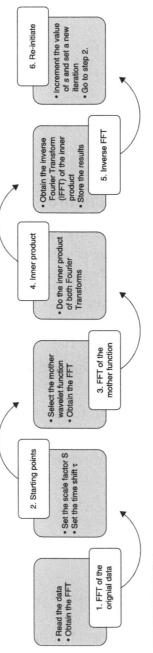

Figure 3.14 Flow diagram of the Wavelet Transform using the convolution theorem

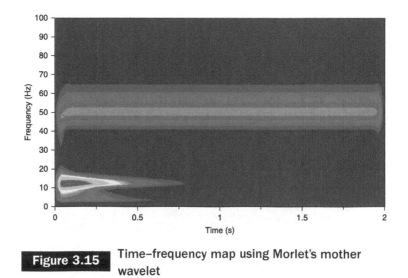

Figure 3.15 Time–frequency map using Morlet's mother wavelet

3.5.5 *Mother wavelet selection*

The selection of the mother wavelet is done considering qualitative and quantitative aspects. For nonlinear vibration analysis and parameter identification, there are three candidates: Gaussian, Paul and Morlet functions. With an appropriate order, the Morlet function gives better results than the other two and it is able to extract the nonlinear behavior in the time–frequency resolution.

Shape mapping is an alternative approach for a qualitative selection; but it is limited to a visual comparison. The quantitative approaches are based on the measurement of the inequality between the original signal and the mother wavelet. There are several methods for measuring this inequality, such as measuring the entropy, calculating the uncertainty or estimating the cross-correlation.

For vibration signals, linear and nonlinear, evaluating the time–frequency resolution is the best criteria for selecting the mother wavelet, especially when analyzing field data.

Figure 3.16 shows the time–frequency map obtained with Paul's mother function while Figure 3.17 shows the same map obtained with the Gaussian Derivative mother function.

In order to determine the parameters of the mother function, it is necessary to set them finding the best time–frequency resolution. In the case of the Gaussian

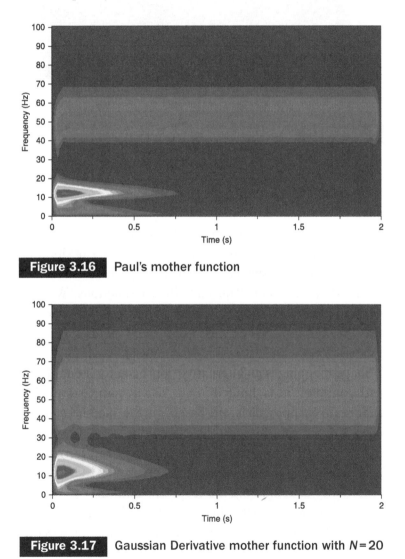

Figure 3.16   Paul's mother function

Figure 3.17   Gaussian Derivative mother function with $N = 20$

Derivative function, lower values of N produces inadequate resolutions, and the wavelet transform can introduce results that differ from the physical phenomenon. For a mechanical system, such as gear boxes, the Morlet mother function can be selected based on the dynamic parameters of the mechanical elements under study. With these features, the nonlinear parameters are easily identified as the frequency responses that vary with respect to time. These features are presented in the next chapter.

3.6 Discrete Wavelet Transform

According to the definition, the scale parameter s and the translation parameter τ vary continuously. Therefore, the CWT of a signal produces redundant information that is useful in signal denoising or feature extraction. These features make CWT a very powerful tool for identifying the parameters of mechanical systems. The drawbacks of the CWT are the computational time and data size.

For other applications, such as image processing or the biomedical process, the CWT gives limited advantages; therefore, it is necessary to have a process for reducing the redundancies without detriment to the information contained in the original signal. This is been done by the application of the Discrete Wavelet Transform (DWT).

The way to implement the DWT is to use a logarithm discretization of the scale factor s and link it to the step size, such that:

$$s = s_0$$
$$\tau = j\tau_0 s_0^n \qquad\qquad [3.31]$$

where $s_0 < 1$, $\tau_0 \neq 0$ and n is an integer number. Then, the family of mother wavelets is expressed as:

79

$$\psi_{j,n}(t) = \frac{1}{\sqrt{s_0^n}} \psi\left(\frac{t - j\tau_0 s_0^n}{s_0^n}\right) \tag{3.32}$$

It is recommended (Pincus 1991; Addison 2002) to use $s_0 = 2$ and $\tau_0 = 1$. The Discrete Wavelet Transform becomes:

$$\langle x|\psi(j,n)\rangle = \frac{1}{\sqrt{2^n}} \int_{-\infty}^{\infty} x(t)\psi^*\left(\frac{t - j2^n}{2^n}\right) dt \tag{3.33}$$

There are some limitations for the development of the discrete mother functions, but they are not presented here, since in this book we only apply the Continuous Wavelet Transform.

There are a lot of mother wavelets available in the internet and in the tool box of Matlab®. The most commonly used are the de Daubechies series, the basis of which is the Haar mother function (D2) which can be expanded up to D20.

	$\psi(t)$
Haar (D2)	1, $\quad 0 \le t < \frac{1}{2}$
	-1, $\quad \frac{1}{2} \le t < 1$
	0, *otherwise*

3.7 Phase diagram

In an autonomous system, the variable t does not appear explicitly in the differential equation of motion. Thus only the differential part will appear. In a dynamic system with a differential equation of the form:

$$\ddot{x} + f(x,\dot{x}) = 0 \tag{3.34}$$

where $f(x,\dot{x})$ represents the nonlinear terms.

If we define the state variable $y=\dot{x}$, then equation 3.34 can be rewritten as a set of two first order differential equations:

$$\dot{x}=y$$
$$\dot{y}=-f(x, y) \qquad [3.35]$$

If x and y represent a Cartesian coordinate system, the plane xy is called the phase plane, state plane or phase diagram. Any pair of coordinates x and y represents a state of the system, and as the states of the system change the phase diagram moves and generates a curve with a specific trajectory. The shape of the trajectory is an indication of the stability of the system.

Equation 3.35 can be rewritten as:

$$\frac{dy}{dt} = y\frac{dy}{dx} = -f(x,y) \qquad [3.36]$$

or

$$\frac{dy}{dx} = -\frac{f(x,y)}{y} = \phi(x,y) \qquad [3.37]$$

Integrating equation 3.37 will be equivalent to the Hamilton's equation

$$H(\bar{q},\bar{p}) = \frac{p^2}{2m} + V \qquad [3.38]$$

If $V = \frac{1}{2}kx^2$ the Hamiltonian will be

$$\frac{1}{2}my^2 + \frac{1}{2}kx^2 = C \qquad [3.39]$$

or

$$y^2 + \omega_n^2 x^2 = C_n \qquad [3.40]$$

which represents an ellipse.

If there is a solution at any point of $\phi(x, y)$, there is a unique slope of the trajectory. If the slopes of the trajectory crosses the x y axes at right angles, then the crossing points are defined as the singular points. Singular points correspond to a state of equilibrium in both velocity and force.

Figure 3.18, shows the phase diagram of equation 3.15 during the transient response. It can be noticed that there are several loops, different amplitudes and different centers. The reason for having different loops is due to the fact that the system has two frequencies, one associated with the transient response (small loops) and the other associated with the steady state response.

Once the transient response vanishes, the phase diagram presents a single ellipse centered at the origin (Figure 3.19). The phase diagram is applied to identify the stability of the system, but it can also be applied for identifying the system's

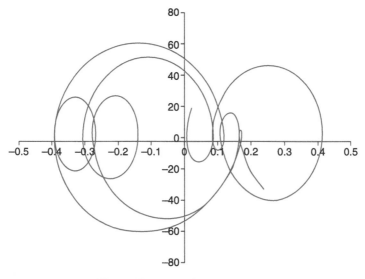

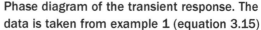

Figure 3.18 Phase diagram of the transient response. The data is taken from example 1 (equation 3.15)

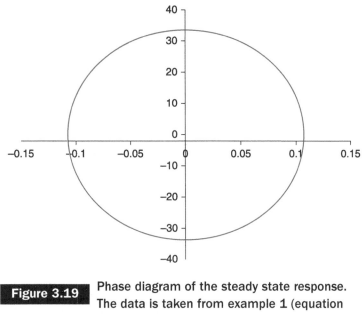

Figure 3.19 Phase diagram of the steady state response. The data is taken from example 1 (equation 3.15)

parameters, since the size of the ellipse depends on knowing the dimensions of the ellipse, at $x=0$, $y=y_{max}$ and at $y=0$, $x=x_{max}$, therefore:

$$\omega_n = \frac{y_{max}}{x_{max}} \qquad [3.41]$$

This equation represents the natural frequency of the system.

Applying equation 3.34 to the data shown in Figure 3.19 we obtained the excitation frequency $\omega_n = 314.1$ *rad/seg*.

3.8 Approximate entropy

The approximate entropy is a technique used to quantify the amount of regularity and the unpredictability of fluctuations over time-series data. This statistical technique can be used for predicting the instabilities of a nonlinear system.

The application of this technique determines 'how' regular a data series is. Thus, it can determine the regularity of vibration data measured from a nonlinear system. A linear system will produce time series of a regular form, whereas nonlinear systems will develope non-regular time series. This technique is described as an algorithm, since it is based on the evaluation of the pattern the time series has. The algorithm is described as follows.

Procedure

Step 1

Divide the data series into a set of N vectors $u(m)$ of dimension m (m represents a time window for comparing the regularity of the time series, and it is a new dimension of the entropy function). The set of vectors are equally spaced in time.

As an example a time series $x(t) = [1, 2, 6, 5, 8, 3, 5]$ will be divided into a set of $N = 6$ vectors of dimension $m = 2$:

$u(1) = [1,2]$

$u(2) = [2,6]$

$u(3) = [6,5]$

$u(4) = [5,8]$

$u(5) = [8,3]$

$u(6) = [3,5]$

Step 2

Select a real positive number r. This value represents the filtering level (which depends on the analysis, but it is recommended to use a value from 10% to 30% of the standard deviation of the original time series).

Step 3

Form a sequence of vectors $v(1)$, $v(2)$, ..., $v(N-m+1)$ where $v(i)=[u(i), u(i+1),..., u(i+m-1)]$. For the previous example

$$v(1)=[u(1), u(2)]$$
$$v(2)=[u(2), u(3)]$$
$$v(3)=[u(3), u(4)]$$
$$v(4)=[u(4), u(5)]$$
$$v(5)=[u(5), u(6)]$$

Step 4

Calculate the distance between all the vectors $v(i)$ as:

$$d[v(i), v(j)]=max|u(k)-u(k^*)|$$

For example

$$d[v(1), v(1)]=max|u(1)-u(2)|$$
$$d[v(1), v(2)]=max|u(1)-u(3)|$$

and so forth.

Then calculate the number of times that the difference is lower than r

$$C_i^m = \frac{(number\ of\ v(j)\ such\ that\ d[v(i),v(j)]<r)}{N-m+1} \qquad [3.42]$$

Step 5

The approximate entropy is defined as:

$$ApEn(m,N,r) = \phi^m(r) - \phi^{m+1}(r)$$

where

$$\phi^m(r) = \frac{\sum_{i=1}^{N-m+1} \log(C_f^{m(r)})}{N-m+1} \qquad [3.43]$$

A detail industrial application is included in the following chapters.

3.9 References

Addison N. (2002) *The Illustrated Wavelet Transform Handbook*, Taylor and Francis, New York.

Cohen, L. (1989) Time frequency distribution – a review, *Proceedings of the IEEE* 77(7), pp. 941–981.

Daubechies, I. (1992) *Ten Lectures on Wavelets*, SIAM, Philadelphia.

Gao, R., and Yan, R. (2011) *Wavelets Theory and Applications for Manufacturing*, Springer, New York.

Gondek, J., Meyer, G., and Newman, J. (1994) 'Wavelength dependent reflectance functions'. In *Computer Graphics, Proceedings, Annual Conference Series*, pp. 213–220.

Mallat, S. (1998) *A Wavelet Tour of Signal Processing*, San Diego, California.

Meyer, Y. (1993) *Wavelets, Algorithms and Applications*, SIAM, Philadelphia.

Pincus, S. M. (1991) 'Approximate entropy as a measure of system complexity'. *Proceedings of the National Academy of Sciences* 88 (6): 2297–2301. doi:10.1073/pnas.88. 6.2297. PMC 51218. PMID 11607165.

Stromberg, J. (1983) 'A modified Franklin System and Higher-Order Spline Systems on R^n as unconditional bases for Hardy space'. *Proceedings of the Conference on Harmonic Analysis in Honor of Antoni Zygmund*, Vol. 2, pp. 475–494.

4

Parameter identification

DOI: 10.1533/9781782421665.87

Abstract: In this chapter, the concepts of parameter identification are discussed. It is clear that parameter identification is a process to determine a mathematical model from experimental data or field measurements. In addition, there are two general forms to develop a model: fitting the data to a function (such as a regression analysis) or constructing a model from a physical representation. The first method is known as nonparametric and the second method is known as parametric. Examples of single and multiple degrees of freedom are presented. Data are taken from actual cases and the procedures are described from field applications. Two types of procedures are included, time domain and frequency domain procedures. For a multiple degrees of freedom application a time–frequency domain procedure is included.

Key words: parameter identification, time domain analysis, frequency analysis.

4.1 Introduction

Engineers and scientists are trying to represent nature's behavior and one way to do this is by representing each

87

phenomenon as an isolated system. Mechanical systems are represented in many different ways, but in general, they can be isolated as a set of punctual masses interconnected by springs and dampers. Although it is a simplistic approach, its application to real problems is sufficient to predict most of the actual behavior. Nevertheless, mechanical systems evolve and nowadays they are faster and lighter than in the past decades. Therefore, linear model, do not always represent their actual behavior.

In this chapter, the concepts of parameter identification are discussed. It is clear that parameter identification is a process to determine a mathematical model from experimental data or field measurements. In addition, there are two general forms to develop a model: fitting the data to a function (such as a regression analysis) or constructing a model from a physical representation. The first method is known as nonparametric and the second method is known as parametric. Nonparametric methods could provide better estimations, but the model itself has no physical meaning. In other words, nonparametric methods are useful for those cases where it is difficult to relate experimental data to a physical phenomenon. Parametric methods assume that a physical model represents the phenomenon, and the model has a mathematical function; then the problem is to estimate the mathematical function from the experimental data.

This chapter describes two parametric methods, and it presents methods for estimating physical models of mechanical systems. The models are obtained analyzing experimental data. These models are represented by a second order differential equation of the form:

$$[M]\ddot{\bar{x}} + [C]\dot{\bar{x}} + [K]\bar{x} = \bar{f}(t) \qquad [4.1]$$

where $[M]$, $[C]$, $[K]$ are matrices representing masses,

damping coefficients and stiffness, and $f(t)$ is a vector containing all the external excitations. This general equation assumes constant coefficients and the challenge is to find values for the three matrices and functions for the external forces. In general, there is a wide range of identification techniques, and selecting the technique depends upon the actual problem. There are three possible approaches:

- time domain analysis
- frequency domain analysis
- nonlinear analysis.

In this chapter, we will present examples of the first two techniques, and in Chapter 5 we will discuss applications of the nonlinear analysis.

In the case of time domain analysis, one should find an approximate analytical expression, written in terms of unknown parameters, and compare it with experimentally measured data. In this case we need to measure only one output signal and then it is possible to perform an identification procedure.

In the case of frequency domain methods, the problem is reduced to find the transfer function of the system and obtain the system's response as a function of the excitation frequency. For these methods, is recommended to excite the system with harmonic functions.

4.2 Time domain analysis

The simplest procedure for determining the coefficients of equation 4.1 is taking advantage of the supper position principle. Applying an impact force to the system is equivalent to having the system's impulse response. For a

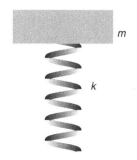

Figure 4.1 Schematic representation of a lumped-mass system

single degree of freedom system, it is equivalent to solving equation 4.1 with an initial condition. Figure 4.1 shows a schematic representation of a lumpedmass system used for determining rubber properties.

A rubber strip that behaves simultaneously as a spring and as a viscous damper suspends the mass. The mass is displaced an initial value x_0 and then it is released. Measuring the mass displacement, the dynamic response can be determined. Figure 4.2 shows the displacement function.

Assuming that the response is harmonic, then the displacement can be approximated as (equation 1.7, Chapter 1)

$$x = e^{pt}[x_0\cos(qt)] \tag{4.2}$$

where

$$p = -\frac{c}{2m} \tag{4.3}$$

and

$$q = \sqrt{\left(\frac{k}{m}\right) - \left(\frac{c}{2m}\right)^2} \tag{4.4}$$

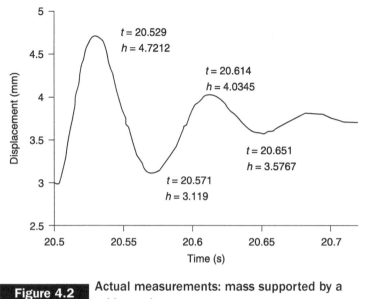

Figure 4.2 Actual measurements: mass supported by a rubber strip

Considering that frequency q depends only on the system stiffness, then

$$q = \sqrt{\left(\frac{k}{m}\right)}$$
[4.5]

Then

$$k = q^2 m$$
[4.6]

Or in terms of the time period

$$k = \left(\frac{2\pi}{T}\right)^2 m$$
[4.7]

Equation 1.4 (Chapter 1) can be used to determine the damping coefficient.

$$\ln\left(\frac{x_1}{x_2}\right) = \frac{c}{2m} T$$
[4.8]

or

$$c = \frac{2m}{T} \ln\left(\frac{x_1}{x_2}\right) \tag{4.9}$$

Using the actual measurements (Figure 4.2) and knowing that the mass weighs $50\,N$:

$$T = 20.614 - 20.529 = 0.085\,s$$

$$\ln\left(\frac{x_1}{x_2}\right) = \ln\left(\frac{4.7212}{4.0345}\right) = 0.1572$$

$$k = 2367.2\;N/m$$
$$c = 18.85\;Ns/m$$

This procedure is quite simple for a single degree of freedom system, but in complex systems it requires special attention. Since the procedure is based on the transfer function, it is necessary to excite the mechanical system with a well-known force. For this purpose, it is recommended to use an instrumented hammer. These type of hammers have an accelerometer mounted inside the hammer which provides a reference force that can trigger the measurement process. Figure 4.3 shows an instrumented hammer with a soft tip, the force is recorded with an accelerometer mounted on the back of the tip.

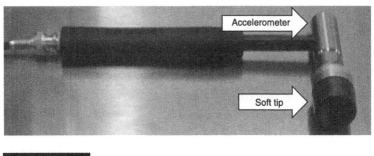

Figure 4.3　Instrumented hammer

The dynamic parameters obtained from an impact test are determined from the transfer function. The hammer's output signal corresponds to the excitation and the transducer signal is the response. In this way, the dynamic parameters are obtained from the transfer function as:

$$H(\Omega) = \frac{reponse}{excitation}$$ [4.10]

where 'response' is the transducer's signal in the frequency domain and 'excitation' is the hammer's signal also in the frequency domain.

$H(\Omega)$ will depend on the type of transducer while the response can be a displacement, a velocity or an acceleration signal. The following table describes the type of response and conversion between the types of responses:

	Displacement	Velocity	Acceleration
$H(\Omega)$	$\dfrac{X}{F}$	$\dfrac{V}{F}$	$\dfrac{A}{F}$
Conversion	1	$i\omega\dfrac{V}{F}$	$-\omega^2\left(\dfrac{A}{F}\right)$

If there is a single mass, the response spectrum will display a dominant frequency; this frequency corresponds to the natural frequency. If there are multiple masses, then the frequency spectrum will display dominant frequencies for several masses. These are the natural frequencies of the dominant degrees of freedom.

4.2.1 Multiple degrees of freedom

Dealing with multiple degrees of freedom systems is not straight forward. Determining individual natural frequencies is somewhat simple if they are clearly visible in a frequency

93

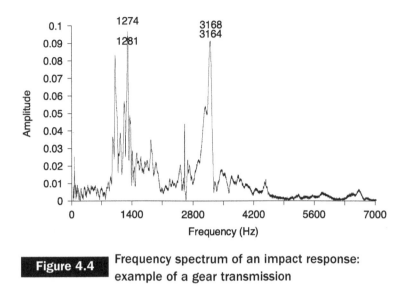

Figure 4.4 Frequency spectrum of an impact response: example of a gear transmission

spectrum, but the corresponding damping coefficients requires a different analysis.

Figure 4.4 shows the frequency spectrum of torsional accelerations of a gear transmission. The excitation force was produced with an instrumented hammer. It can be noticed that there are two dominant frequencies corresponding to each degree of freedom (input and output shafts). There is also a coupling frequency that can be identified in a time–frequency map. Torsional stiffness are approximated as:

$$\omega_{ni}^2 = \frac{K_i}{J_i} \qquad [4.11]$$

The inertias are estimated from the transmission design, and, for this example, the following table gives the torsional stiffness:

Shaft	Frequency (Hz)	Inertia (kg-m^2)	Stiffness (kg-m)
1	1274	1.345×10^{-3}	8.61×10^4
2	3164	5.718×10^{-3}	2.26×10^6

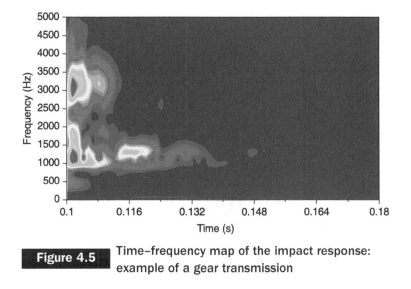

Figure 4.5 Time–frequency map of the impact response: example of a gear transmission

It is almost impossible to determine the damping coefficients from a sole analysis of the frequency spectrum, particularly when real and imaginary spectrums are similar. A way to overcome this difficulty is by analyzing the output response in a time–frequency domain (Figure 4.5). The construction of a time–frequency map was presented in Chapter 3.

The map shows the behavior of the two dominant frequencies. It is clear that they decrease differently in time. While frequency 1 (1274 Hz) dies out in 0.14 s, frequency 2 (3164 Hz) vanishes in 0.115 s. If we analyze only the behavior of each frequency, it is possible to approximate the time variation as an exponential function. Figure 4.6 shows the time variation of the output response only for frequency 1. It is evident how the amplitude decreases and we can assume that this decrement can be associated to the damping coefficient of mass 1. Similarly, Figure 4.7 shows amplitude variations of frequency 2. For both cases, the damping

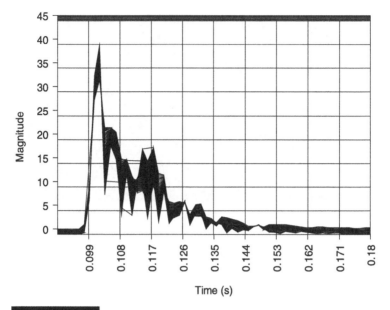

Figure 4.6 Time variation of frequency response 1

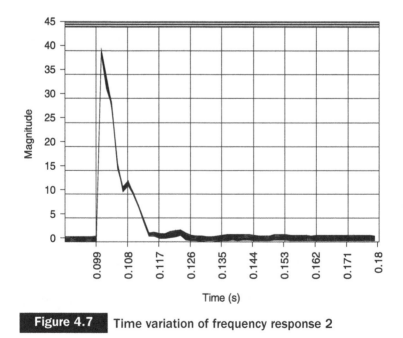

Figure 4.7 Time variation of frequency response 2

coefficients can be obtained applying equation 4.9. The following table is a summary for this example:

Shaft	X_1	X_2	T (s)	Damping coefficient
1	40	2	4.5×10^{-2}	8.67×10^{2}
2	40	2	1.7×10^{-2}	2.93×10^{3}

From practical experience, it is known that the coupling frequency is lower than the other two frequencies. In this case, the coupling parameters (stiffness and damping) occur at 480 Hz. Figure 4.8 shows time variations of the coupling frequency. Following the same procedure, the coupling stiffness and damping parameters are listed in the following

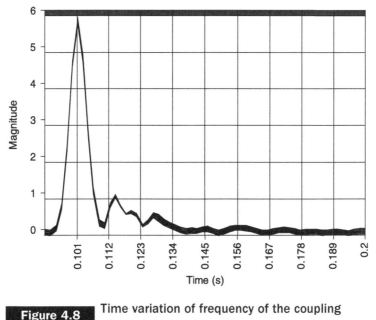

Figure 4.8 Time variation of frequency of the coupling element

table. The stiffness is calculated from the coupling frequency and the logarithmic decrement is calculated with:

$$T = 0.033\,s$$
$$X_1 = 6$$
$$X_2 = 0.5$$

Thus, the corresponding values are:

Coupling	Stiffness	Damping coefficient
1	1.22×10^4	2.02×10^{-1}
2	5.20×10^4	8.61×10^{-1}

Finally, an approximated model will have the following equation (Chapter 1):

$$J_1\ddot{\theta}_1 + C_1\dot{\theta}_1 + C_{12}(\dot{\theta}_1 - \dot{\theta}_2) + K_1\theta_1 + K_{12}(\theta_1 - \theta_2) = 0$$
$$J_2 2 + C_2\dot{\theta}_2 + C_{21}(\dot{\theta}_2 - \dot{\theta}_1) + K_2\theta_2 + K_{21}(\theta_2 - \theta_1) = 0$$

[4.12]

Substituting the calculated values:

$$1.345x10^{-3}\ddot{\theta}_1 + 8.67x10^2\dot{\theta}_1 + 2.02x10^{-1}\,(\dot{\theta}_1 - \dot{\theta}_2)$$
$$+ 8.61x10^4\theta_1 + 1.22x10^4(\theta_1 - \theta_2) = 0$$

$$5.718x10^{-3}\ddot{\theta}_2 + 2.93x10^2\dot{\theta}_2 + 8.61x10^{-1}\,(\dot{\theta}_2 - \dot{\theta}_1)$$
$$+ 2.26x10^4\theta_2 + 5.20x10^4(\theta_2 - \theta_1) = 0$$

Time domain analyses are suitable for impact tests and they give very good results. The accuracy of the test can be improved by repeating the measurements two or three times and conducting statistical analyses of the resulting values. This procedure gives different results depending on the type of support. If the system can be suspended, then it is possible to isolate the foundation effects and estimate real parameters. In many cases, this is not possible and the system should be tested in real conditions. Therefore, it is important to identify the foundation stiffness.

4.3 Frequency domain analysis

Parameter identification is essential to construct the dynamic model of any mechanical system. With this model, it is possible to reproduce its dynamic behavior, evaluate modifications on a machine or validate new designs. There are many applications of parameter: rotor bearings, gear transmissions, machining process, etc.

The estimation of dynamic parameters is possible because data acquisition systems can collect large amounts of data at high speed, allowing the analysis of dynamic signals of a high frequency domain. A typical acquisition system collects data at 2 MS/s (Mega samples per second) and has a resolution 16 bits per data.

Time domain processes are very sensitive to the test procedure, and it is easier to determine incorrect parameters. To overcome this difficulty frequency domain processes are widely used. These processes are able to identify systems with a finite number of generalized coordinates q_i.

In order to apply frequency domain procedures, it is necessary to have, additionally to vibration sensors, an excitation system with variable frequency, and the frequency range should be large enough to excite some of the natural frequencies of the system. The procedure assumes that the dynamic system could be represented as a set of linear equations of the form

$$[M]\{\ddot{q}_i\} + [C]\{\dot{q}_i\} + [K]\{q_i\} = \{f_i\} \qquad [4.13]$$

Since the system is assumed to be linear, q_i will have small amplitudes and the matrices are considered to be constant.

It is important to relate $[M]$, $[C]$ and $[K]$ to physical quantities of the actual system. If the system is represented as a second differential equation, it is fundamental to set the experimental conditions such that the amplitudes are always small. In this way, the data will be reliable and repeatable. The other assumption is that the coefficients are frequency independent.

A system's parameters are estimated because it is almost impossible to directly measure them. Frequency domain methods are based on estimating the transfer function between a known force (external excitation) and a vibration measurement (displacement, velocity of acceleration). The transfer function for a single degree of freedom system has the form (Chapter 1)

$$|T| = H(\Omega) = \frac{\sqrt{1 - \left(\frac{2\xi\Omega}{\omega_n}\right)^2}}{\sqrt{\left(1 - \left(\frac{\Omega}{\omega_n}\right)^2\right)^2 + \left(\frac{2\xi\Omega}{\omega_n}\right)^2}} \qquad [4.14]$$

The excitation force should be a function of the excitation frequency Ω. The system should be able to control the excitation force within a wide range of frequencies.

Procedure

1. Apply a force $\{f_1\}$ and record $\{q_1\}$ and $\{q_2\}$ (q_1 is orthogonal to q_2)
2. Apply a second force $\{f_2\}$ and record $\{q_1\}$ and $\{q_2\}$
3. Transform $\{f_1\}, \{q_1\}, \{f_2\}$ and $\{q_2\}$ into the frequency domain

$$F_1 = \int_{-\infty}^{\infty} \{f_1\} e^{i\Omega\tau} \, d\tau$$
$$F_2 = \int_{-\infty}^{\infty} \{f_2\} e^{i\Omega\tau} \, d\tau$$
$$Q_1 = \int_{-\infty}^{\infty} \{q_1\} e^{i\Omega\tau} \, d\tau \qquad [4.15]$$
$$Q_2 = \int_{-\infty}^{\infty} \{q_2\} e^{i\Omega\tau} \, d\tau$$

If the measurement data are velocities or accelerations, then the displacement vectors can be obtain as

$$i\Omega Q_1 = \int_{-\infty}^{\infty} \{\dot{q}_i\} e^{i\Omega\tau} d\tau$$

$$-\Omega^2 Q_1 = \int_{-\infty}^{\infty} \{\ddot{q}_i\} e^{i\Omega\tau} d\tau$$

[4.16]

Substituting equations 4.15 and 4.16 into equation 4.13 a new set of linear equations is obtained

$$-[M]\Omega^2 Q_1 + i\Omega[C]Q_1 + [K]Q_1 = F_1$$
$$-[M]\Omega^2 Q_2 + i\Omega[C]Q_2 + [K]Q_2 = F_2$$

[4.17]

Or

$$[H]Q_1 = F_1$$
$$[H]Q_2 = F_2$$

[4.18]

where

$$[H] = -[M]\Omega^2 + [K] + i\Omega[C]$$

[4.19]

Rearranging equations 4.17, and defining Q_{11} as the displacement q_1 due to F_1, Q_{12} as the displacement q_1 due to F_2, Q_{21} as the displacement q_2 due to F_1, and Q_{22} as the displacement q_2 due to F_2, then

$$[H] \begin{bmatrix} Q_{11} & Q_{12} \\ Q_{21} & Q_{22} \end{bmatrix} = \begin{Bmatrix} F_1 \\ F_2 \end{Bmatrix}$$

[4.20]

$[H]$ can be obtained from equation 4.19, it will be a set of complex numbers containing the system's parameters. The excitation forces must be significantly different in order to have two different vectors Q_1 and Q_2.

Applying a set of harmonic excitation forces will produce a set of measurement vectors. Recording the measurements at each corresponding frequency will produce enough data to estimate the system's parameters. A preliminary approximation can be obtained with

$$Real[H] = -[M]\Omega^2 + [K]$$

[4.21]

$$Imag[H] = \Omega[C]$$

[4.22]

The next example illustrates this method.

Example

A two degrees of freedom system, such as the example of Chapter 1, was excited with two independent harmonic forces. The model representing the system has the form

$$\ddot{x}_1 + \frac{k_1}{m_1}x_1 + \frac{k_e}{m_1}[(x_1 - x_2)] = \frac{P_1(t)}{m_1}$$

$$\ddot{x}_2 + \frac{k_2}{m_2}x_2 - \frac{k_e}{m_2}[(x_1 - x_2)] = \frac{P_2(t)}{m_2} \qquad [4.23]$$

As it was discussed previously, it is recommended to excite the system near the natural frequencies. For this example, the displacements of m_1 and m_2 were stored in vectors x_1 and x_2 The natural frequencies were determined with an impact test and determined from a frequency spectrum (Figure 4.9). From this figure, the natural frequencies are

$$\omega_1 = 146.5 \, \text{Hz}$$

$$\omega_2 = 103.8 \, \text{Hz}$$

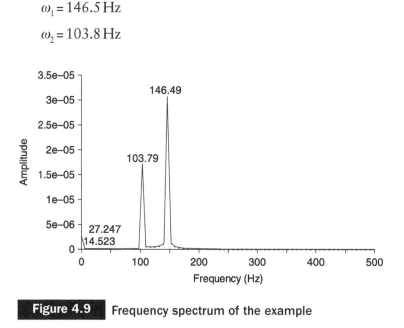

Figure 4.9 Frequency spectrum of the example

The first run consisted of applying P_1 to mass m_1 from $17\,Hz$ up to $160\,Hz$ and storing x_1 and x_2. Then the data was transformed into the frequency domain using the Fast Fourier Transform (FFT equation 4.15). Then, a second force P_2 was applied to mass m_2 and similarly the data was stored and transformed with the FFT.

At each frequency, a four by four matrix was obtained and, ideally, this matrix should be symmetrical. The diagonal values correspond to the main stiffness k_1, k_2 and the other two values are related to the coupling stiffness k_e.

If we plot the real values of matrix H as a function of Ω, we obtain curves similar to those presented in Figures 4.10–4.12:

$$\text{Real}[H] = \begin{bmatrix} H_{11} & H_{12} \\ H_{21} & H_{22} \end{bmatrix} \qquad [4.24]$$

In order to determine the parameters of the system, it is necessary to fit a curve to each of the plots and find the intersection to zero; the corresponding frequency will be ω_0. From equation 4.10 it is easy to find that the natural frequency will occur when

$$H_{ij} = 0 = k_{ij} - \omega_0 m_{ij} \qquad [4.25]$$

Or

$$\frac{k_{ij}}{m_{ij}} = \omega_0$$

For this example, the results are

For H_{11} the corresponding frequency is $\omega_{01} = 150\,Hz$
For H_{22} the corresponding frequency is $\omega_{02} = 105\,Hz$
The average H_{12} is 64.

Other techniques increase the accuracy of the method such as the system transfer functions. This method is based on

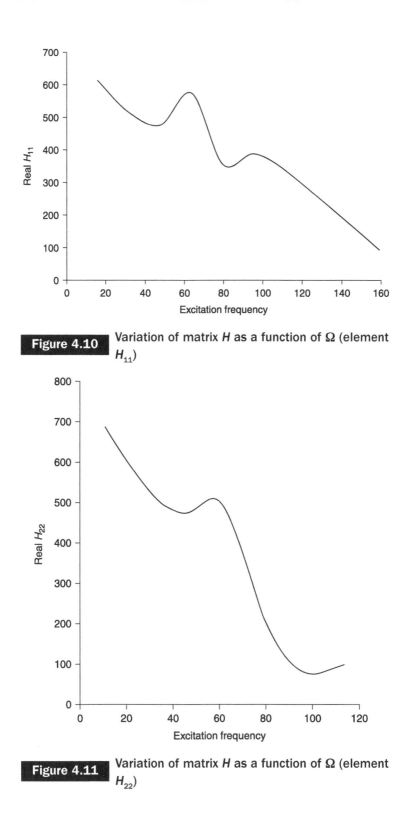

Figure 4.10 Variation of matrix H as a function of Ω (element H_{11})

Figure 4.11 Variation of matrix H as a function of Ω (element H_{22})

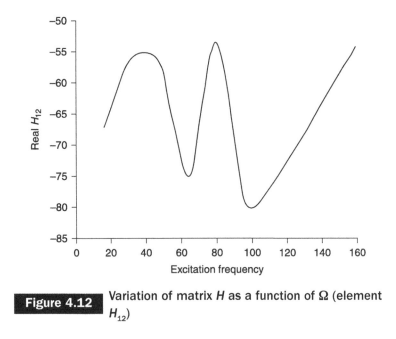

Figure 4.12 Variation of matrix *H* as a function of Ω (element H_{12})

calculating the inverse of matrix *H*. Another method is the Instrumental Variable Filter method. Both methods use the inverse matrix as a weighting function in a minimization procedure. Using this minimization procedure leads to more accurate results, but the process requires a complex minimization scheme. The discussion of these methods is beyond the scope of this chapter.

5

Application of signal processing to mechanical systems

DOI: 10.1533/9781782421665.107

Abstract: Nonlinear responses are difficult to identify with conventional monitoring systems; the wave-form plus the frequency variations can be interpreted as noise or random vibrations, but they have a particular pattern that can be identified with other techniques. Every machine element produces a characteristic response that allows its identification. In this chapter, results from nonlinear models of mechanical systems are analyzed with different signal processes. These analyses serve as references for the analysis of field data obtained from different machinery. Examples include machine elements as bearings, gears and components under direct friction. The data were analyzed with the Fourier Transform, the Continuous Wavelet Transform and the phase diagram.

Key words: nonlinear elements, bearings, gears, friction, gear boxes.

5.1 Introduction

In this chapter, the nonlinear dynamic behavior of different mechanical systems is introduced. Previous chapters described the complexity of the identification of nonlinear parameters. For this reason, it is important to understand, from dynamic models, the relationship between the nonlinear parameters and their dynamic response.

The analysis of vibration measurements is the basis for monitoring mechanical systems and machinery. These monitoring systems are based on the spectrum analysis of time series data obtained from accelerometers, or other transducers, but they are limited to the analysis of the excitation frequencies. When the nonlinear behavior of the system is stronger than the excitation frequencies, the analysis is very complex and, in many cases, the frequency of the response changes depending on the amplitude of the applied force. This is the reason why we use other analysis techniques in order to identify the nonlinear response.

Monitoring systems predict the remaining life of much industrial machinery; they are also suitable for rotating machinery. The life prediction is based on the tendency analysis of the excitation frequencies produced by each of the rotating elements. The selection of the frequencies is very straightforward and its implementation assures good results when the amplitudes are significantly high. The limitation of these systems is that they cannot predict failures in early stages. When a failure occurs, it starts developing vibrations of low amplitude which excites the nonlinear modes.

Nonlinear responses are difficult to identify with conventional monitoring systems; the wave form, plus the frequency variations can be interpreted as noise or random vibrations, but they have a particular pattern that can be identified with other techniques. Every machine element produces a characteristic response that allows its identification. In this way, the identification of the nonlinear response can be applied for the prediction of the element life, and it can be set such that early failures are clearly identified. The main elements that have strong nonlinearities are roller bearings and gears. Other phenomena that behave nonlinearly are friction and rubbing, and they are also discussed in this chapter.

5.2 Roller bearings

From the beginning, humans have moved objects over the surface, and one of the challenges is the reduction of friction. There are many solutions to this problem, but the most practical has been to slide the object over roller elements. Based on this simple principle, roller bearings were created and they are widely used; therefore, it is necessary to understand their dynamic behavior. From a kinematic analysis, it is possible to see that they are a complex mechanical system with strong nonlinearities that will be described next.

A typical ball-bearing system consists of five contact parts: the shaft, the inner ring, the rolling elements, outer ring and the housing (Figure 5.1). The deformation of each part will influence the load distribution and, in turn, the service life of the bearing. It is well known that classical calculation methods cannot accurately predict load distributions inside the bearing. The best representation of the stiffness of the roller bearing is analyzing the deformation of the elements with the Hertz contact stress formula.

Ball bearings (Figure 5.2) are very stiff compared with sliding bearings; their stiffness can be approximated by a set

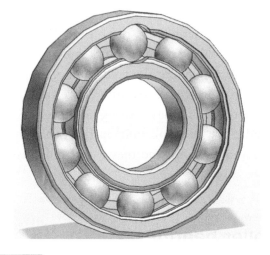

Figure 5.1 Elements of a roller bearing

Figure 5.2 Schematic representation of the nonlinear stiffness of the rollers

of individual springs, where the number of springs supporting the shaft varies with the angular position of the shaft. This variation depends upon the kinematics of the ball or roller as it moves around the shaft. Thus, the ratio of rotation of the roller as a function of shaft's rotation is determined from its motion: it rotates around its geometric centroid and it translates around the shaft. The kinematics of a roller inside a bearing is described in Figure 5.3.

The translation movement produces two epicyclical motions, one relative to the inner track and the other relative to the outer track. The spinning of the roller is calculated as:

$$\omega_b = \frac{D\omega_s}{d} \qquad\qquad [5.1]$$

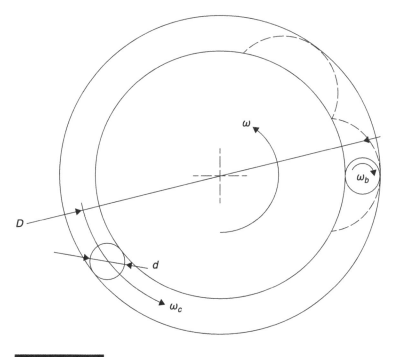

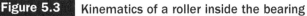

Figure 5.3   Kinematics of a roller inside the bearing

The translation of the roller changes the angular position of each roller element, the angular translation is found as:

$$\phi = \frac{d \cos(\omega_b t)}{D + \dfrac{d}{2}} \cos(\alpha) \qquad\qquad [5.2]$$

where D is the diameter of the trajectory of the rollers, d is the roller diameter and α is the axial angle (Figure 5.4).

The angular position of a roller element can be determined from Figure 5.5.

The number of rollers in contact depends on the preloading of the bearing; this preloading deforms the rollers and determines the number of rollers in contact. Figure 5.5 shows a schematic representation of the kinematics of the roller's deformation. The nonlinear characteristic of the rolling bearing is the roller-track deformation. Based on this idea, the stiffness of the roller is determined from the roller-track deformation and it is calculated with the Hertz equation. Since the roller translates around the shaft, the number of balls supporting the load varies with the angular position of

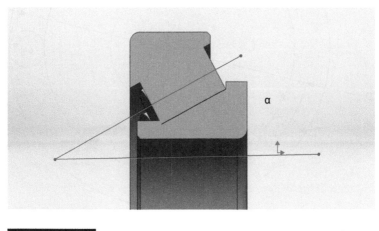

Figure 5.4 Axial angle

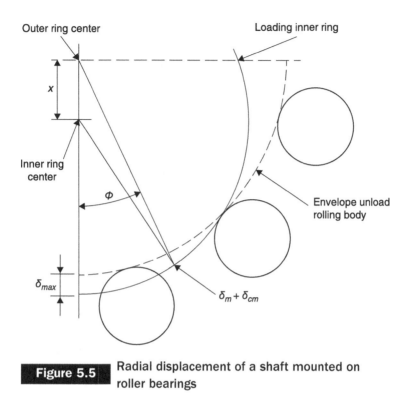

Outer ring center

Loading inner ring

x

Inner ring center

Φ

Envelope unload rolling body

δ_{max}

$\delta_m + \delta_{cm}$

Figure 5.5 Radial displacement of a shaft mounted on roller bearings

the shaft; this translation effect modifies the overall stiffness of the bearing. Although this variation may be small, it creates a nonlinear vibration, which turns out to be relatively difficult to identify in field problems.

The stiffness of the track–roller interaction is a function of the deformation and it can be calculated as:

$$P_i = E \sqrt{\frac{Dd}{D-d} \left(\frac{\delta_i}{\alpha}\right)^3} \qquad [5.3]$$

where P_i is the normal force and δ_i is the deformation of an individual roller (according to Figure 5.5). From this figure, it can be noticed that the nonlinear effect comes from the relative deformation of the rollers as they roll around the shaft and the number of rollers supporting the load. The individual deformation depends upon the angular position or the roller element and it can be calculated with:

$$\delta_i = \delta_{max} \cos\left[\frac{\pi}{2}i\left(\phi + \frac{2(i-1)}{N}\right)\right]$$ [5.4]

where N is the number of rollers.

The total force due to the deformation of the rollers inside the track will be:

$$P = \Sigma P_i$$ [5.5]

The solution of the dynamic model requires the definition of the transmitted force $F(t)$. Ideally, it should be constant, and equal to the radial force, but this is not the case. First of all, the radial force varies according to every application, and the rolling bearing itself produces a specific type of excitation force. Rolling bearings generates transient vibrations due to stiffness nonlinearities and structural defects. The external defects can be associated with the kinematics of the roller. Therefore, there are four external sources of vibration; two of them are associated with the angular velocity of the ball ω_b and their angular translation ω_ϕ. The other two frequencies are related to structural defects on the inner and outer tracks. These external frequencies excite the nonlinear terms which makes difficult its analysis with the Fourier Transform. When a bearing is new, the amplitude of the excitation forces are due to manufacturing deviations (these deviations are very small and produce low amplitude vibrations). In a frequency spectrum their amplitude values are under most of the other sources of vibration. The types of defects that change the amplitude of the excitation forces are, in general, pitting marks on the tracks or on the roller, cracks or excessive wear.

The representation of the external excitation forces can be represented as:

$$F(t) = F_{ir} \cos(\omega_{ir}) + F_{or} \cos(\omega_{or}) + F_c \cos(\omega_c) + F_{re} \cos(\omega_{re}) \qquad [5.6]$$

The four frequencies shown in equation 5.6 are calculated with the following equations:

- Contact frequency between the roller element and the internal track ω_{ir}:

$$\omega_{ir} = \frac{N}{2}\left[1 + \frac{d}{D}\cos(\alpha)\right]\omega \qquad\qquad [5.7]$$

- Contact frequency between the roller element and the external track ω_{or}:

$$\omega_{or} = \frac{N}{2}\left[1 - \frac{d}{D}\cos(\alpha)\right]\omega \qquad\qquad [5.8]$$

- The casing frequency ω_c:

$$\omega_c = \frac{1}{2}\left[1 - \frac{d}{D}\cos(\alpha)\right]\omega \qquad\qquad [5.9]$$

- The roller spin frequency ω_{re}:

$$\omega_{re} = \frac{D}{d}\left[1 - \left(\frac{d}{D}\cos(\alpha)\right)^2\right]\omega \qquad\qquad [5.10]$$

The four factors multiplying the rotation speed ω are non-integer numbers and they produce non-synchronous vibrations.

The rolling bearing can be modeled as a mass-spring system (Figure 5.6). This simple model represents the behavior of a shaft mounted on roller bearings; it is accurate enough to identify the nonlinear effects.

The dynamic equation is found considering that the mass moves along the direction of the external force, therefore only a one degree of freedom model is needed. The damping (c) effect is caused by the structural damping.

$$m\ddot{y} + c\dot{y} + P = me\omega^2 \cos(\omega t) + F(t) \qquad\qquad [5.11]$$

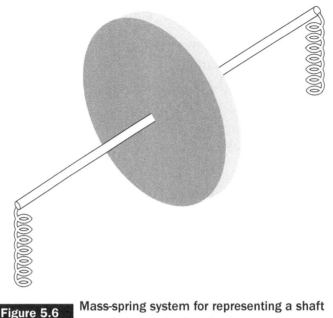

Figure 5.6   Mass-spring system for representing a shaft mounted on roller bearings

where $me\omega^2$ is the shaft's imbalance. The solution is found using the Runge-Kutta method.

One of the challenges of a monitoring system is the identification of early faults in rolling bearings. As we have described before, failures in bearings start at surface level; thus, they generate a relatively small energy vibration compared to other sources, and its identification is very cumbersome. With the application of phase diagram plots and time–frequency maps, early failures can be predicted in real time. The process is as follows.

Vibrations are measured with a transducer, preferably an accelerometer. Then, the signal is analogically integrated in real time. Then, in an oscilloscope, the phase diagram is plotted. When the equipment is new, the first diagram corresponds to the healthy reference. Since we know that bearings have a nonlinear response, and that this response is the result of its stiffness dependency on frequency, we can monitor the phase diagram in order to "see" the moment when instabilities occur. In this way, if we permanently monitor the "shape" of the phase diagram, and we detect the appearance of instabilities, then we will be able to detect early faults.

In a time–frequency map the nonlinear effect will display gradient variations along the time axis; these variations will be displayed at the excitation frequencies or around the linear natural frequency.

The model described in equation 5.11 was solved for each of the excitation frequencies. The first solution is the free vibration solution where all the excitation frequencies are neglected as if the bearing were new (healthy bearing). This condition will show the nonlinear effects, and will display the frequency variations along the time axis.

Figure 5.7 shows the phase diagram, the frequency spectrum and the time–frequency map of a healthy bearing. The phase diagram is calculated using the Hamilton's principle, the frequency spectrum is found with the Fast Fourier Transform, and the time–frequency map is produced with the Continuous Wavelet Transform, using Morlet's mother function with eight significant terms. The shape of the phase diagram shows the effect of the nonlinear stiffness of the bearing (the ovoid corresponds to the stiffness changes as a function of the frequency). The frequency spectrum shows two significant frequencies, one at 2 Hz and another at 28 Hz. These frequencies are associated with the mass-spring movement, but their time variations are not displayed. The time–frequency map (contour plot) displays the two main frequencies, but also shows their variations along the time axis. It is clear that the 28 Hz frequency changes periodically, and these variations confirm the shape changes displayed in the phase diagram.

The following analysis corresponds to the case when the excitation is due to the imbalance. Figure 5.8 shows the

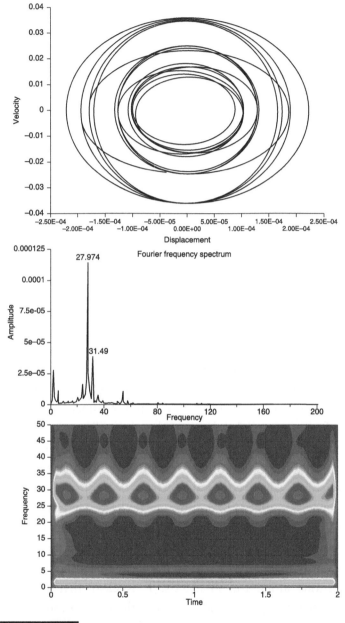

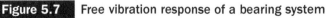

Figure 5.7 Free vibration response of a bearing system

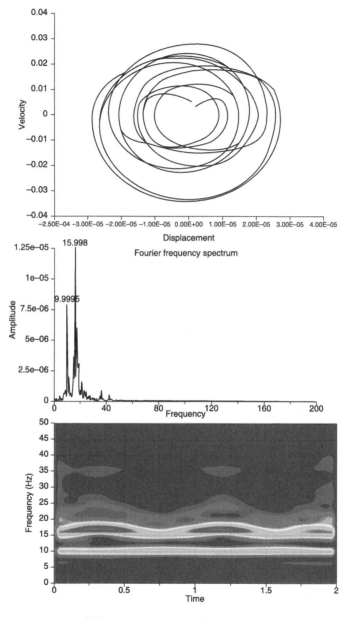

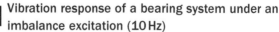
Vibration response of a bearing system under an imbalance excitation (10 Hz)

phase diagram, the frequency spectrum and the time–frequency map. As in the free vibration analysis, it clearly identifies the nonlinear effects. The excitation frequency was set at 10 Hz, with this value. The Fourier spectrum shows the highest amplitude at almost 16 Hz. The spectrum hardly shows the same values as the previous one. For this reason, the analysis is based on the phase diagram and the time–frequency map. The phase diagram shows a different pattern compared with the free vibration case. The shape changes in magnitude and it has an apple shape, rather than an elliptic shape. The time–frequency map gives more information regarding the nonlinear response. The 28 Hz frequency (from the free vibration response) has very low amplitude. The dominant frequency varies along the time axis and has a harmonic frequency at 32 Hz, with the same time variation.

If the excitation frequency is higher (30 Hz) the response is totally different. It was expected that the amplitude would change, but in this case the shape of the response also changes. The excitation frequency is similar to the nonlinear frequency identified with the free vibration simulation. In this case, the shape of the phase diagram is different: instead of having an apple shape, it has a lemon shape. It also shows variations in the amplitude that are due to the nonlinearities. The frequency spectrum shows two significant picks, one at 21 Hz and the other at 30 Hz. It is unclear what the first pick is related to. The time–frequency map gives better information regarding the nature of the system. It shows that the higher amplitude occurs around 28 Hz (similar to the free vibration response). It also shows that this response varies with respect to time, and this variation is periodical. The periodical variations have a frequency of around 10 Hz. It can be considered that this is the "time–frequency signature" of an imbalance shaft mounted on healthy bearings.

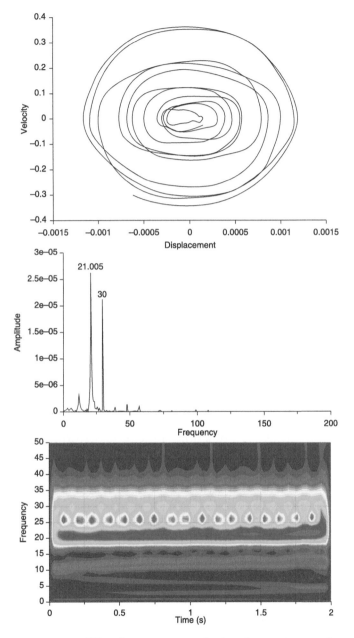

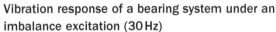

Figure 5.9 Vibration response of a bearing system under an imbalance excitation (30 Hz)

Since the stiffness of the bearing depends on the displacement, it is important to analyze the effect of the excitation amplitude on the vibration response. Figure 5.10 shows a similar analysis considering a higher amplitude excitation. The instability of the response is identified in the three diagrams. The phase diagram has a lemon shape, but is larger diameter rotates from the vertical axis to the horizontal axis. The Fourier spectrum shows a "noisy" behavior around 1500 Hz. And the time–frequency map displays two dominant frequencies, one around 900 Hz and the other around 1500 Hz. Both frequencies change with time and have periodic response along the time axis. With this analysis, it is clear that the system response changes drastically as a function of the excitation frequency and the amplitude of the excitation. Since the stiffness depends on the amplitude and the rotation of the shaft, the frequency response changes as a function of the excitation. This is a valuable concept that can help in the analysis of field data.

It is important to notice that the nonlinear effect depends upon the excitation frequency and the excitation amplitude. In general, the amplitude dominates the nonlinear response since the stiffness is a function of the roller deformation, and this deformation depends on the external force.

In order to simulate a damaged bearing, different types of failures were represented as excitation forces; each force has a frequency equivalent to the type of failure. If a failure occurs at the roller surface, it is simulated as an excitation force with a frequency equal to the roller frequency (equation 5.10). Or, if the failure occurs at the inner ring, then it is simulated as an excitation force acting at the inner ring frequency (equation 5.7). Figure 5.11 represents the same three analyses (phase diagram, frequency spectrum and time–frequency map) for a simulation considering a failure at the roller. The time–frequency map shows variations in

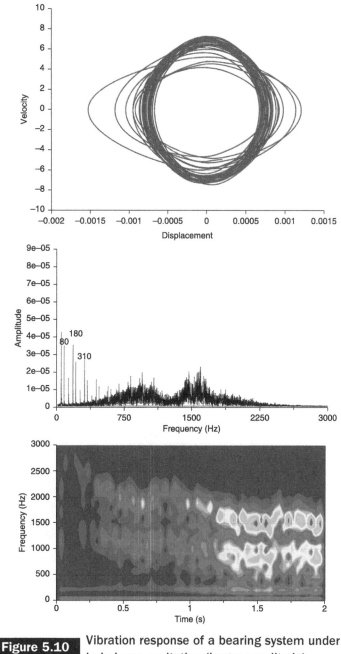

Figure 5.10 Vibration response of a bearing system under an imbalance excitation (larger amplitude)

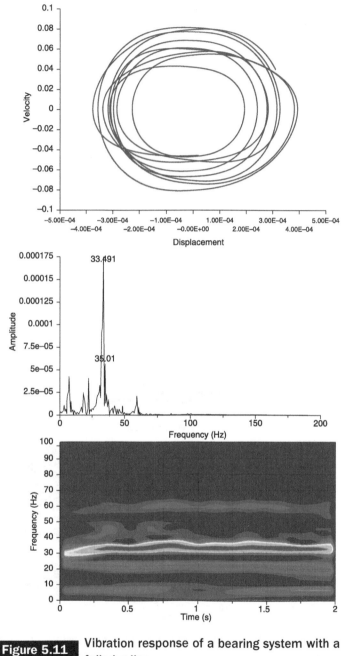

Figure 5.11 Vibration response of a bearing system with a failed roller

the frequency response around 60 Hz and the phase diagram describes how the response depends upon the rotation speed; the shape of the diagram is oval instead of elliptical and the center of the figure "moves" away from the origin along the horizontal axis.

When a failure occurs in the roller's cage, the dynamic behavior presents a different response. Figure 5.12 shows three analyses of the simulation of a cage failure. The dominant frequency appears around 31 Hz and at this frequency the amplitude changes with time. The phase diagram describes the nonlinear behavior; it is clear how the vibration amplitude changes from a low amplitude shape (small loop) to a larger loop. The shape of the loops is irregular and the center moves along the displacement axis. These characteristics are clear indications of the nonlinear response.

Inner ring and outer ring failures have higher frequencies than the shaft's fundamental frequency (rotating frequency). For this reason the dynamic response differs from the other type of failures. Figure 5.13 shows the same three analyses for a situation where a failure occurs at the inner ring. The phase diagram shows a condition where several loops appear around a major loop; even more, the center of the loop moves from a positive value to a negative value. The time–frequency map describes three dominant frequencies, one at low frequency, one around 31 Hz, and its corresponding first harmonic. The shape of the dominant frequency displays variations at steady periods along the time axis.

If a failure occurs at the outer ring, then an excitation force will occur every time that a roller element passes over the failure. Figure 5.14 shows the response of a system simulating an outer ring failure. In this case, the phase diagram shows a different pattern compared to an inner ring failure. Although the shape has less loops, it shows a non-homogeneous pattern and the orbits circulate around two centers, one at the right

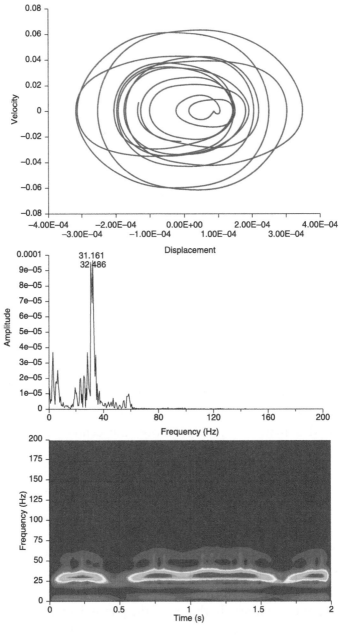

 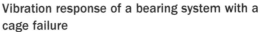 Vibration response of a bearing system with a cage failure

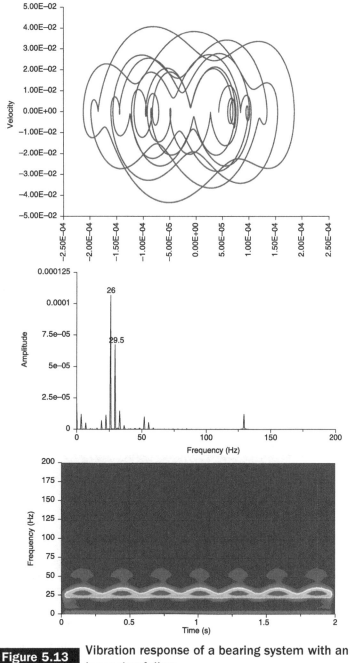

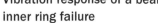 Vibration response of a bearing system with an inner ring failure

127

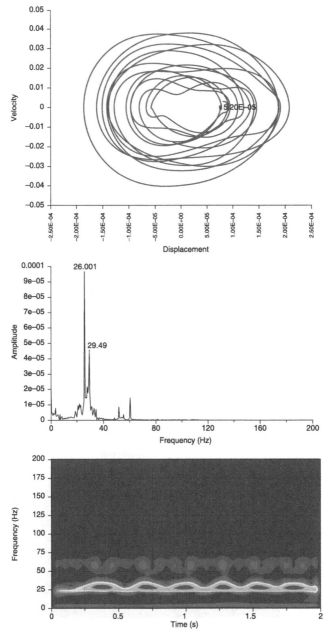

 Vibration response of a bearing system with an outer ring failure

position, and the other at the left position (relative to $x = 0$). On the other hand, the time–frequency map shows a dominant frequency at 31 Hz, with steady variations along the time axis. These variations are similar to those displayed in the previous case.

Finally, the case where all the excitation frequencies occur is presented in Figure 5.15. The system's behavior is identified at the phase diagram and the time–frequency map. The frequency spectrum shows a similar pattern as those corresponding to individual failures; therefore, it is difficult to identify the vibration source. The phase diagram displays the nonlinear behavior and, in this case, two attracting poles appear again (there are two centers along the displacement axis: one at a positive position and the other at a negative position). The time–frequency map shows that at 31 Hz there are variations along the time axis, but also around the frequency axis. These variations are due to the change in the bearing stiffness which depends upon the load amplitude. The excitation force that has a larger impact in the stiffness is the unbalanced force; meanwhile the other excitations forces affect the frequency response.

Monitoring the phase diagram can predict the instance when a significant variation will occur. The main distinction for this analysis is the fact that the nonlinear response will produce two attracting poles along the displacement axis; thus, if the excitation frequency changes, the shape in the diagram will have a single center, but if the force amplitude changes, then the stiffness will change and the response will display two centers or attracting poles. From the time–frequency map it can be concluded that it is necessary to monitor other frequencies rather than only monitoring the excitation frequencies. When a failure occurs at the rings, higher frequencies appear and they produce a sub-synchronous response with a lower frequency.

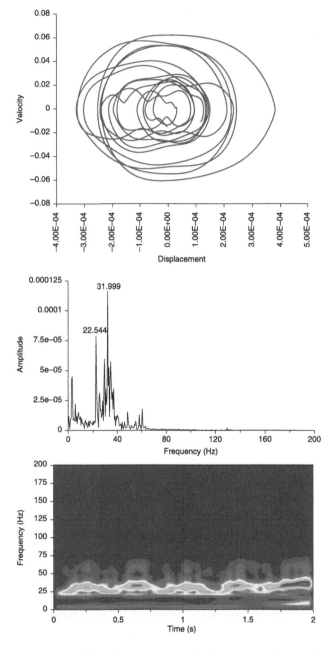

 Vibration response of a bearing system with all excitation forces acting simultaneously

5.3 Gears

Gears are mechanical elements widely applied in adjusting rotational speed. They have been studied for several decades and there are many specialized publications regarding their design, manufacturing, dynamics and other related topics. Nevertheless, they represent a particular field of interest in mechanical vibrations. This interest comes from the nature of the relative motion between the matching teeth. Ideally, a pair of matching teeth has an involute profile that assures a continuous motion; thus, the relative velocity of the two wheels is constant. In practice, this relative velocity varies during the gearing action due to several factors: manufacturing errors, elastic deformations, and variations on the gear mesh stiffness. This effect is relevant to the nonlinear behavior and is the core of the following section. The difference between the ideal constant velocity ratio and the actual motion is defined as the transmission error; it is referred to as the deviation from a perfect motion between the driver and the driven gears. And it is the combination of different gear variations, such as non-perfect tooth profile, pitch errors, elastic deformations, backlash, etc. It is also the dominant cause of gear noise and vibrations. The simplest type of noise is a steady note which may have a harmonic content at gear mesh frequency. This frequency is normally modulated by the rotating frequency. Modulated noise is often described as a buzzing sound. In general, gears show a frequency-modulated spectrum with a distinguished mesh frequency and side band space at the shaft rotating frequency. Other noises are associated with pitch errors. They are described as scrunching, grating, grouching, etc. They contain a wide range of frequencies that are a lot higher than the rotating frequency. White noise can also be present and it may be associated with loss of contact between the teeth.

5.3.1 Gear mesh stiffness

Gear stiffness is determined from the teeth bending deformation. The deformation depends on the number of teeth in contact, which changes as the gear rotates. Thus, the stiffness is a function of the displacement of the teeth. The representation of the actual stiffness is complicated and can be represented with different functions, but the simplest approximation is a piecewise function. This approximation takes into consideration the main nonlinear effect; other terms are neglected because they have no significant contribution to the present analysis and their contribution to the vibration signal is insignificant to actual vibration sensors.

The stiffness of a tooth pair can be represented as the combination of the individual elastic bending deformation of each tooth in contact. The simplest representation of a tooth deformation is:

$$\sigma = \frac{Mc}{I} = \frac{6W_t L}{bh^2} \qquad [5.12]$$

where b is the face width, W_t is the transmitted force, L is the height of Lewis' parabola and h its width. Considering the beam's maximum deflection δ:

$$\delta = \frac{W_t L^3}{3EI} \qquad [5.13]$$

Thus, tooth stiffness can be approximated as:

$$S_t = \frac{W_t}{\delta} = \frac{3EI}{L^3} = \frac{1}{4}Eb\left(\frac{h}{L}\right)^3 \qquad [5.14]$$

From Figure 5.17, it is clear that $\tan(\alpha/2) = h/(2L)$, and in most case $\alpha = 30°$, then:

$$S_t = 2Eb\left(\tan\left(\frac{\alpha}{2}\right)\right)^3 \qquad [5.15]$$

Combining the stiffness of the teeth in action, and assuming that the stiffness depends upon the contact ratio, the mathematic expression of the piecewise stiffness function is:

$$
S_t = \left\{ \begin{array}{l} 2Eb\tan\left(\dfrac{\alpha}{2}\right)^3 (n+1), 0 < \theta < \dfrac{2\pi}{N}\left(m_p - n\right) \\[4mm] 2Eb\tan\left(\dfrac{\alpha}{2}\right)^3 (n+1), \dfrac{2\pi}{N}\left(m_p - n\right) < 0 < \dfrac{2\pi}{N} \end{array} \right\}
\qquad [5.16]
$$

In this equation, N is the number of teeth, n is the minimum teeth in contact, m_p is the contact ratio, E is the elastic modulus, and α is calculated from Figure 5.16.

Contact ratio can be derived from a spur gear action, and then extrapolated to other geometries. Its demonstration is out of the scope of this book, but the general formula is:

$$
m_p = \frac{\sqrt{(R_{OBP}^2 - R_{bP}^2)} + \sqrt{(R_{OBG}^2 - R_{bG}^2)} - C_O \sin(\theta_O)}{P_b}
\qquad [5.17]
$$

Contact between teeth takes place along the line of action; that is, the line where matching gears are in contact, limited by the outside diameters of both gears (Figure 5.16). The time of conjugated motion is determined from the time when a tooth starts contacting the matting tooth until it misses contact. This conjugated motion takes place along the line of action and the number of teeth in contact is determined from the gear action. Gear action is divided in two: from the beginning of contact until they reach the pitch diameter, and from the pitch diameter until the end of contact. The sliding velocity along these sections is different. In the first case it is negative, thus the oil film is reduced, whereas in the second case the sliding velocity is positive and it brings more oil to the contact area. It is remarkable that the variation in oil film causes variations in the damping coefficient as the teeth transmit motion. During gear action, more than one pair of teeth is in contact. The contact

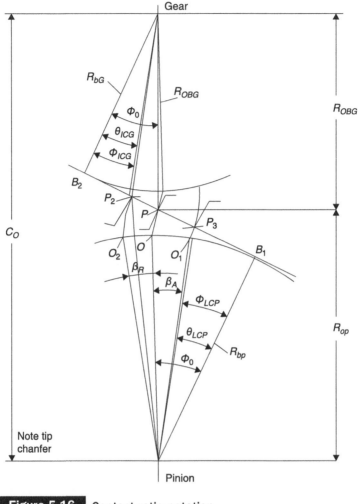

Figure 5.16 Contact ratio notation

ratio defines the percentage of time that more than one pair of teeth are in contact. For example, if the gear ratio is 1.43, it means that 43% of the time there are two pairs of gears in contact and 57% of the time only one pair of teeth transmit the torque. This is the dominant effect on the gear mesh stiffness.

In most gear pair systems, torsional motion is coupled by the gear pair stiffness; therefore a two degree of freedom

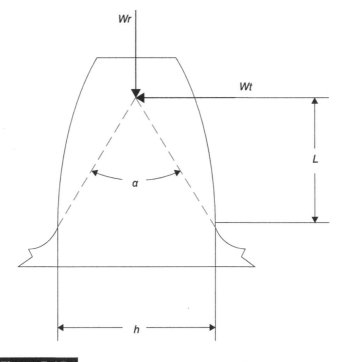

Figure 5.17 Gear tooth geometry and stiffness parameters

model will reflect accurately most practical applications. If it is necessary to include other effects, increasing the degrees of freedom could accommodate other compliances that are present in the system.

Many researches and engineers have developed a significant number of gear dynamic models. Most of them have been developed for the prediction of noise and vibrations, and they have demonstrated that gear vibrations are highly nonlinear. In this chapter we presented one of the most commonly used models that is widely accepted. It has been demonstrated that this simplified lumped-mass model is adequate for predicting torsional vibrations in a gear box. There are many representations of this system, but it can identify most of the common vibration and noise problems.

The only ones that it cannot reproduce are those associated with backlash. In the case of gear boxes with several speed reductions and light casings, more sophisticated models are needed. On the other hand, this simplified model illustrates how a gear reduction can be analyzed as a system with stiffness discontinuities, and it allows us the identification of this characteristic in the vibration response.

Figure 5.18 shows an idealized stiffness function. This idealization reflects the nonlinear behavior of a pair of teeth, including other terms which complicate the solution, and it gives no better insight of the dynamic behavior. By expanding equation 5.16 as a Fourier series, this function can be approximated as a polynomial equation of the form:

$$S_t = 2Eb\tan\left(\frac{\alpha}{2}\right)^3 (n+1)[c_0 + c_1\theta + c_2\theta^2 + c_3\theta^3] \qquad [5.18]$$

where

$$c_0 = \left[m_p + \sum_i \frac{1}{\pi i}\sin\left(2\pi i(m_p - n)\right)\right] \qquad [5.19]$$

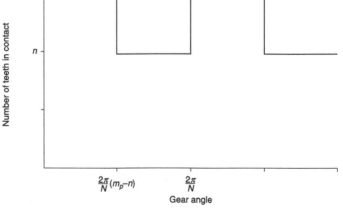

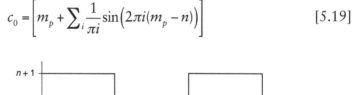

Figure 5.18 Gear teeth stiffness as a function of the angular rotation

$$c_1 = \sum_i \frac{N}{\pi i}\left[1 - \cos\left(2\pi i(m_p - n)\right)\right] \qquad [5.20]$$

$$c_2 = \sum_i \frac{N^2 i}{\pi}\left[\sin\left(2\pi i(m_p - n)\right)\right] \qquad [5.21]$$

$$c_3 = \sum_i \frac{N^3 i^2}{6\pi}\left[1 - \cos\left(2\pi i(m_p - n)\right)\right] \qquad [5.22]$$

N is the pinion teeth number, m_p is the gear contact ratio and n is the integer part of the contact ratio.

5.3.2 Gear's dynamic model

With this approximation, the dynamic model can be simplified as a 4DOF (four degrees of freedom) lumped-mass model. The model is compounded of two masses m_1 and m_2 (pinion, gear and their shafts) with their corresponding rotational inertias J_1 and J_2, pitch radius r_1 and r_2, lateral stiffness S_1 and S_2, teeth stiffness S_t, and gear force P_e. The independent coordinate system and the generalized displacements are x_1, x_2, θ_1, θ_2, which correspond to two translations along the line of action and the twist of each shaft.

From Figure 5.19 the equation of motion can be represented as a four nonlinear differential equations. Applying Lagrange's principle, the equations of motion are:

$$\ddot{x}_1 + \frac{S_1}{m_1}x_1 + \frac{S_t}{m_1}\left[(x_1 - x_2) + (r_1\theta_1 + r_2\theta_2)\right] = \frac{P_1(t)}{m_1}$$

$$\ddot{x}_2 + \frac{S_2}{m_2}x_2 - \frac{S_t}{m_2}\left[(x_1 - x_2) + (r_1\theta_1 + r_2\theta_2)\right] = \frac{P_2(t)}{m_2}$$

$$\ddot{\theta}_1 + \frac{S_t r_1}{J_1}\left[(x_1 - x_2) + (r_1\theta_1 + r_2\theta_2)\right] = \frac{r_1 P_e(t)}{J_1}$$

$$\ddot{\theta}_2 - \frac{S_t r_2}{J_2}\left[(x_1 - x_2) + (r_1\theta_1 + r_2\theta_2)\right] = \frac{-r_2 P_e(t)}{J_2} \qquad [5.23]$$

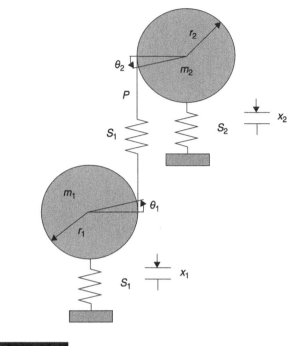

Figure 5.19 Four degrees of freedom lumped-mass model

Along the line of action, the external excitations are included in $P_1(t)$ and $P_2(t)$. These dynamic effects are caused by the gear rotation. Gear teeth defects excite the system at two frequencies: one is the rotating frequency, and the other is the gear mesh frequency. The external gear force has at least these terms:

- Run out effect (F_r) is a sinusoidal function with a 2π period and amplitude of the gear eccentricity.

- Pitch error (f_p) is a short wave length function with a period of $2\pi/N$, and the accumulated pitch effort (F_p) is a long wave function with a period of 2π.

- The profile angle error is another source of vibration and it can be represented as a long period function: $F_\varphi = \varepsilon\tan(\omega t)$.

Typical values for an automobile gear are (mm):

F_r	f_p	F_p	F_φ
0.045	0.02	0.063	0.032

Other effects can also be included but in general they are present only in special cases that are out of the scope of this chapter. Assuming that the geometric errors affect the teeth displacement, the external forces are expressed as:

$$P_e(t) = S_t[(F_p + F_r)sen(\Omega t) + (f_p + F_\varphi)sen(N\Omega t)] \qquad [5.24]$$

where Ω is the pinion's rotational speed.

Equations 5.23 and 5.24 include the nonlinear effects caused by the gear action. In Chapter 1, the linear solution was presented and it was shown that the natural frequencies of the system can be calculated finding the eigenvalues of the system. In this case, there is no closed form solution and the dynamic response is found numerically. The most stable solution method is the Runge-Kutta algorithm. With this model, the gear faults can be identified since it incorporates most of the significant effects. It is important to know that there are two types of vibrations in a gear box. One type comes from the external forces, and has a well described form (typically a modulated signal): long wave signal at gear speed frequency over a short wave signal at a frequency equals the speed frequency times the number of teeth (known as gear frequency). These two signals are easily identified with any vibration measuring system. Accuracy manufacturing is the only way of reducing the effects of this type of vibration. The nonlinear response can be minimized during the design process, where the design criterion should be reducing the transmission error. The simplest solution is keeping the total contact ratio as close to an integer number as possible. The other aspect that can be controlled at the design stage is minimizing the elastic deformation and assuring a truly involute shape when the teeth are deformed. This procedure is useful for applications with low torque variations;

but in general, transmissions operate in a wide range of torque conditions and the optimum teeth form is difficult to achieve.

Gears also have some damping, caused by friction forces during teeth contact, and they are determined using modal damping. On the other hand, friction depends on the relative velocity during gear action and on the oil film thickness. The sliding velocity is calculated as:

$$v_{sp} = (\rho_p \omega_p - \rho_g \omega_g) \qquad [5.25]$$

where $\rho_p \omega_p$ are the rolling radius and rotational speed for the pinion and $\rho_g \omega_g$ are the rolling radius and rotational speed for the gear. The rolling radius is calculated as:

$$\rho_g = R_{bp} \tan(\alpha) \qquad [5.26]$$

where α is the pressure angle at the radius of contact. It is important to remark that the pressure angle varies along the line of action, and its reference value is defined at pitch diameter. Thus, the sliding velocity depends upon the position of the teeth along the line of action.

It is clear that, even for the simplest gear model, a gear box will develop chaotic and periodic motions. The quadratic and cubic terms of the gear mesh stiffness are relatively high compared to the linear terms; then, the system can be analyzed as a Duffing's system.

Example
The following example is based on the same parameters as the example in Chapter 1. The following figures show the response under different loading.

	Gear 1	Gear 2
Mass (*m*)	4.87 kg	10.57 kg
Moment of inertia (*J*)	0.0011 kg-m^2	0.004652 kg-m^2
Base radius (*r*)	0.031853 m	0.0483285 m
Bearing stiffness (*K*)	3.66×10^6 N-m	4.1410^6 N-m
Gear mesh stiffness (*Ke*)		4.37×10^5 N-m

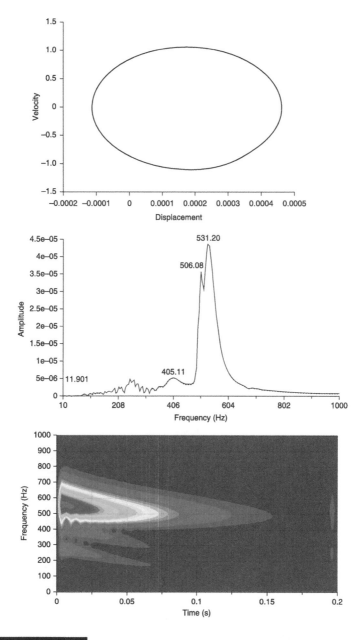

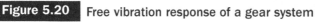

Figure 5.20 Free vibration response of a gear system

141

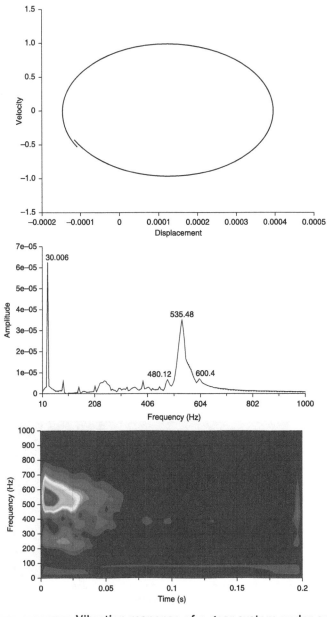

 Vibration response of a gear system under an unbalanced force rotating at 30 Hz

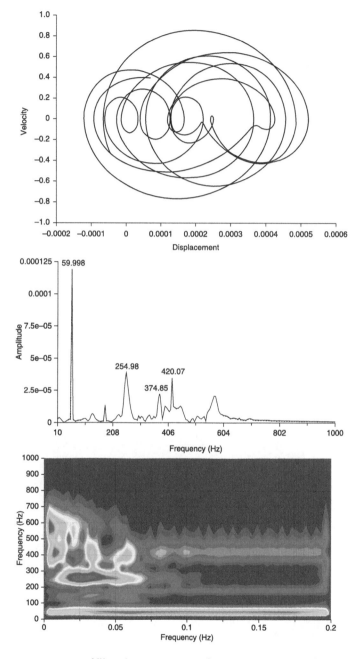

Figure 5.22 Vibration response of a gear system under an unbalanced force rotating at 60 Hz

143

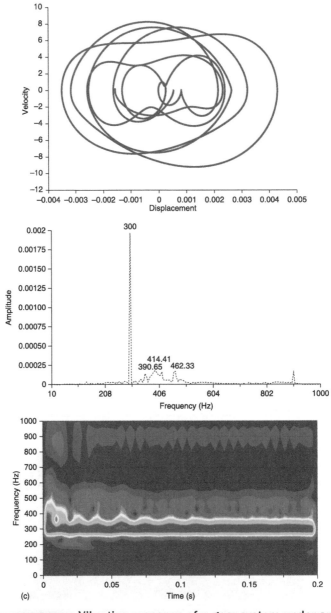

 Vibration response of a gear system under an unbalanced force rotating at 300 Hz

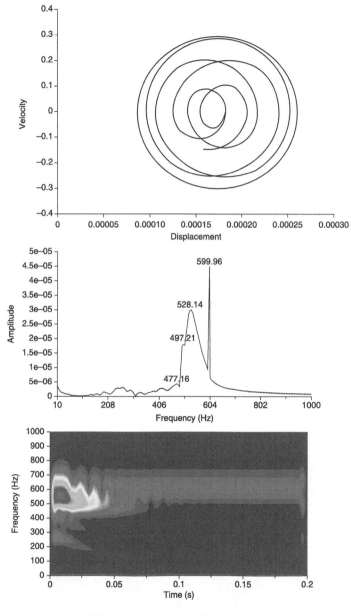

Figure 5.24 Vibration response of a gear system simulating a teeth defect (30 Hz)

145

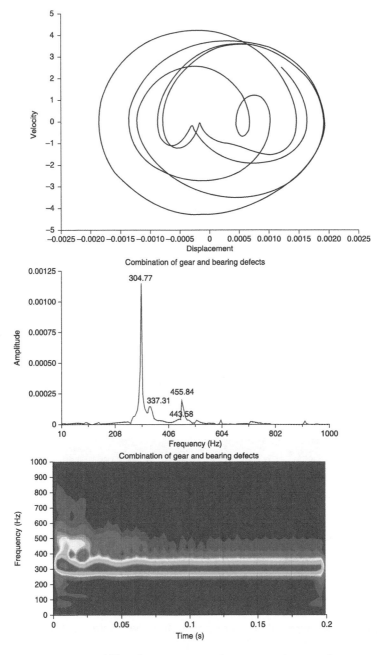

Figure 5.25 Vibration response of a gear system under an unbalanced force rotating at 30 Hz

5.4 Friction

Friction is a phenomenon present in any mechanical system. Its behavior is not fully understood and it draws the interest of many researchers. The dynamic response of a mechanical system with friction between its elements is highly nonlinear. This nonlinearity comes from the friction forces acting between the elements, and these forces vary depending on the direction of the force, the displacement and velocity amplitudes of the relative motion and the shape of the surface. There are several theories that describe friction; in general, it is assumed that the force is proportional to a friction coefficient as described in Figure 5.26.

There are two types of friction coefficient: static and kinetic coefficients. Distinctions between coefficients of static and kinetic friction have been mentioned in the literature for

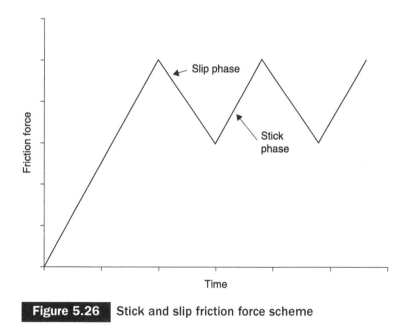

Figure 5.26 Stick and slip friction force scheme

centuries. Euler developed a mechanical model to explain the origins of frictional resistance. He arrived at the conclusion that friction during sliding motion should be smaller. The experiment proposed by Euler involved the sliding of a body down an inclined plane at slopes close to the critical slope at which sliding initiates. This, of course, would mean that, as soon as sliding initiates, a drop of friction force occurs, the difference between static and kinetic friction forces being responsible for the acceleration of the body down the inclined plane.

The distinction between static and kinetic friction was also a major topic of Coulomb's detailed experimental study. Coulomb's work is, in addition, the first major reference dealing with the increase of the coefficient of static friction with increasing times of repose (stationary contact before the initiation of sliding). Coulomb observed a dependence of the kinetic friction on the sliding velocity and a dependence of the static friction on the time of repose. However, for dry metal-to-metal interfaces all those distinctions or variations were absent or negligible.

In general, the coefficient of kinetic friction would be small but increasing with sliding velocity at low velocities. Then, at some velocity (dependent on the materials and the normal pressure) it would achieve a maximum value after which it would decrease with the increase of speed. But also such variations were smaller for the case of dry interfaces. For sufficiently high sliding speeds it is clear that decreases of kinetic friction with increasing speeds may indeed occur in the dry sliding of metallic bodies. However, for the small, the situation is not so clear. Conflicting results obtained with various lubrication conditions and the absence of a clear understanding on the distinction between dry and lubricated sliding added much to the confusion, the question being: Is the dominant effect the static or the kinetic term?

The sliding process is not a continuous one; the motion proceeds by jerks. The metallic surfaces "stick" together until, as a result of the gradually increasing pull, there is a sudden break with a consequent very rapid "slip". The surfaces stick again and the process is repeated indefinitely. When the surfaces are of the same metal, the behavior is somewhat different. Large fluctuations in the friction still occur but they are comparatively slow and very irregular. The average value of the frictional force is considerably higher than that found for dissimilar metals and a well-marked and characteristic track is formed during the sliding.

It has also been observed that the frequency of the stick-slip motion increases with the increase of the driving velocity and that the maximum value of this frequency approaches the undamped natural frequency of the system, although in some cases the oscillation stops at a level well below that natural frequency.

Friction is an important source of damping in many mechanical systems. In fact, in systems such as turbomachinery rotors and large flexible space structures, dry friction may be the most important source of energy dissipation. The viscous-like damping property suggests that many mechanical designs can be improved by configuring frictional interfaces in ways that allow normal forces to vary with displacement. In some applications, classic dry friction is inadequate to suppress vibration. The main difficulty is that it can be strongly nonlinear, and the dissipation effect could produce other undesirable effects.

Friction force has been defined with different models. A well accepted model is known as Martin's model. His model considers only a single coefficient and he modifies the amplitude as a function of the speed. From Figure 5.26, the friction force depends on the vertical displacement and the horizontal velocity as:

$$F_{\mu} = \mu(y^2)sgn(\dot{x} - v) \qquad\qquad [5.27]$$

It is assumed that the block moves at a constant speed v and it oscillates in both directions. This model is relatively simple, but it gives good approximation in many applications. But in other cases, it is necessary to include more elaborated models.

As will be shown, a system with amplitude-dependent friction is more likely to experience intermittent sticking. If the system sticks a significant amount of time, the energy dissipation capability may be seriously degraded. Hence, special care is taken in this analysis to examine sticking conditions (in the case of gear teeth action, sticky occurs only for very high contact stresses). In general, sticking can occur only when the sliding velocity is zero. For such cases, a general form of the friction force can be represented as:

$$F_{\mu} = \mu(C_0 + C_1|x| + C_2|\dot{x}|)sgn(\dot{x}) \qquad\qquad [5.28]$$

where x represents the sliding displacement, $\dot{x}$ represents the sliding velocity, C_0 is the normal force, C_1 is the friction interface amplitude, C_2 is the friction interface velocity and μ is the coefficient of friction (in general it is equivalent to the static coefficient of friction).

The dynamic model for a single degree of freedom system is represented as:

$$m\ddot{x} + c\dot{x} + kx + F_{\mu} = F_e\cos(\omega t) \qquad\qquad [5.29]$$

The system is positively damped at all times and it is clearly stable in the sense of Lyapunov. However, the system is not asymptotically stable for $C_0 \neq 0$. This condition is identified from the phase diagram when $F_e = 0$ and the initial $x = 0.01$.

The following figures represent the solution of equation 5.29. Figure 5.27 shows the phase diagram, where it can be noted that the solution is stable but the shape shows the slip-stick effect. The buckle has two shapes, a time when the

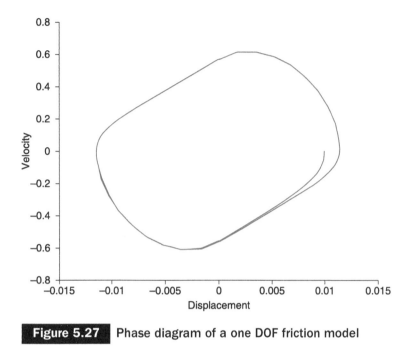

Figure 5.27 Phase diagram of a one DOF friction model

displacement and velocity are linearly related and then a period when they have variable shape. Regardless of the initial condition, the friction force stabilizes the response and, similar to the Van der Pol equation, the solution remains in the same loop.

Figure 5.28 shows the acceleration response. It shows the presence of two jerks per cycle, which correspond to the change in shape in the phase diagram.

The frequency spectrum is shown in Figure 5.29 while Figure 5.30 shows the time–frequency map. Since the response is stable, the response only shows horizontal strips along the time axis.

One of the situations when friction plays a critical role is in rotor dynamics. When a rotor rubs the casing or the journal hits the bearing two additional effects take place. On one hand, the stiffness increases due to the contact between

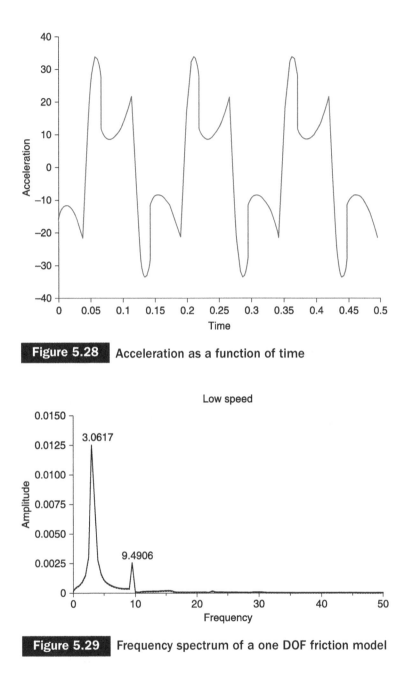

Figure 5.28 Acceleration as a function of time

Figure 5.29 Frequency spectrum of a one DOF friction model

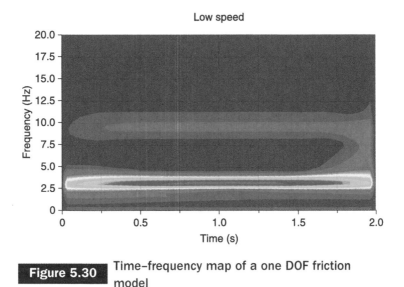

Figure 5.30 Time–frequency map of a one DOF friction model

the rotor and the casing. On the other hand, there is a tangential force due to the friction effect. This situation is illustrated in Figure 5.31.

The equation of motion is derived from Figure 5.30. Using a Newton–Euler notation, the differential equations are:

$$m\ddot{x} + c_x\dot{x} + k_x x = F_x + m\rho\Omega^2 \cos(\varphi_0 + \Omega t) \qquad [5.30]$$

$$m\ddot{y} + c_y y + k_y y = F_y + m\rho\Omega^2 \sin(\varphi_0 + \Omega t)$$

where the stiffness and damping coefficients correspond to the rotor dynamics and the friction effect is included into the external forces as

$$F_x = -F_N \cos(\phi) + F_T \sin(\phi) \qquad [5.31]$$
$$F_x = -F_N \cos(\phi) + F_T \sin(\phi)$$

In this way, the friction effect is included into the tangential force F_T and the stiffness effect is included into the normal force F_N

$$F_T = \mu(C_0 + C_1|s| + C_2|\dot{s}|)sgn(\dot{s}) \qquad [5.32]$$

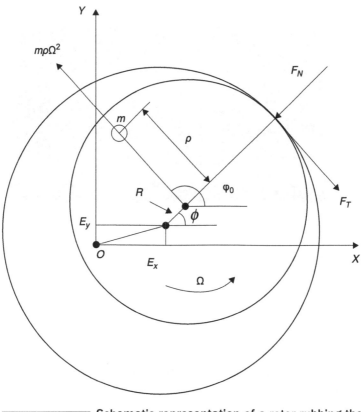

Figure 5.31 Schematic representation of a rotor rubbing the casing

and

$$F_N = \begin{cases} 0, & R \le c \\ K_\mu(R-c), & R \ge c \end{cases} \qquad [5.33]$$

s is a vector perpendicular to R such that

$$s_x = R\sin(\phi) \qquad [5.34]$$
$$s_y = -R\cos(\phi)$$

and the total displacement can be calculated as:

$$x = E_x + s_x \qquad [5.35]$$

$$y = E_y + s_y$$

K_μ can be determined using Hert's contact stress formula and it would be a function of the angular position ϕ.

Although the two DOF model is a simple approximation, its numerical solution is cumbersome and it depends on the friction coefficients and the stiffness function. As an example, Figure 5.32 represents a common frequency spectrum of a rotor rubbing the casing. (The spectrum corresponds to real data obtained experimentally, where a single mass rotor rubbed a wall and it rotated at 60 rps.)

Since the spectrum shows a wide range of frequencies, the time–frequency map was calculated in two ranges, the first range includes a low frequency range (0–2000 Hz). This map (Figure 5.33) shows the nonlinear effect at frequencies above 1000 Hz. At higher frequencies the map shows only nonlinear effects (Figure 5.34).

From the results presented before, it is clear that friction is a complex dynamic problem. Its theoretical formulation and physical analysis give an idea of the dynamic response.

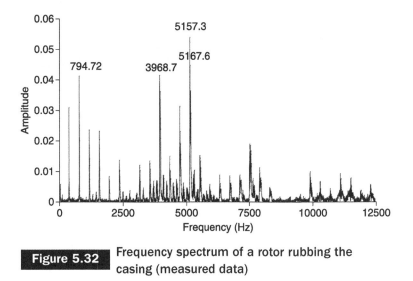

Figure 5.32 Frequency spectrum of a rotor rubbing the casing (measured data)

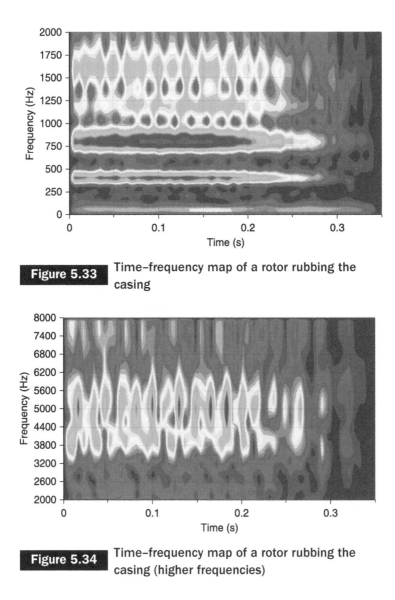

Figure 5.33 Time–frequency map of a rotor rubbing the casing

Figure 5.34 Time–frequency map of a rotor rubbing the casing (higher frequencies)

The analysis of nonlinear vibrations can overcome the limitations of actual vibration monitoring systems. In the three examples presented before, the sole analysis of the frequency spectrum is insufficient to determine the system response. In each case, the phase diagram (plane diagram)

and the time–frequency map give "signatures" of each phenomenon. These signatures are references for the analysis of similar systems. The big challenge, as presented in Chapter 4, is to formulate a mathematical model able to represent the actual behavior. And, although the model could predict the behavior, the estimation of the model parameters is a cumbersome task. In the next chapter some industrial applications will be presented.

6

Practical experience and industrial applications

DOI: 10.1533/9781782421665.159

Abstract: Vibrations are present everywhere, especially in industrial machinery where different types of equipment are connected, and they transmit motion and energy among them. Modern machinery yields more nonlinear vibrations than former equipment and this situation requires new techniques for their analysis. In previous chapters we presented the background for nonlinear vibrations, analytical results of theoretical models and the application of signal analysis techniques to identify nonlinear behaviors. In this chapter, similar procedures are applied to field data. Data were obtained from industrial equipment and vibration measurements were recorded from experimental devices. Results show good agreement with numerical results, and these examples serve as guidelines to other practical applications.

Key words: field data, simulation, gear boxes.

6.1 Introduction

In previous chapters, the theoretical background needed for identifying the dynamic behavior of a mechanical system was described. These concepts were introduced to help engineers in the analysis and prediction of vibration problems in general, and some common cases were discussed in Chapter 5. Nevertheless, applying these concepts to practical cases is not simple. The main difficulty comes from their complexity, since they consist of many elements interconnected by elastic components. For simplicity, each element is considered as a single mass, with various degrees of freedom, and each connecting component could have a linear or nonlinear behavior.

Vibrations are present everywhere, especially in industrial machinery where different types of equipment are connected, and they transmit motion and energy among them. In general, those systems are analyzed or evaluated considering a linear behavior, and vibration monitoring is conducted using the Fast Fourier Transform. In this type of application, all the measurements are associated to a particular excitation force, and from the spectrum analysis each frequency corresponds to a vibration source. Nevertheless this analysis is very useful to predict and monitor machinery conditions ('health condition'); but, it is insufficient if a simulation or a deeper analysis is needed.

In order to illustrate all the concepts presented in previous chapters, a detail industrial application is presented. In this case, the example is a gear transmission, since they are built with several mechanical elements deploying nonlinear behavior, such as gears and rolling bearings. Also they can be considered as a complex system due to the number of rotating elements and total number of degrees of freedom. A linear analysis is relatively simple, it can be done with a linear lumped-mass model, such as the one presented in Chapter 1, or it can be done using finite element models. With these two methods, it is difficult to calibrate the models since they will deliver data different from actual measurements. Even though the instrumentation is accurate enough, analyzing output data requires special attention. This chapter is divided into two sections, the first section describes the test procedure and the analysis of the measurements, and the second section describes the simulation analysis and a comparison of the results.

6.2 Test procedure

First of all, it is important to define the scope of the analysis. Due to the system's complexity, vibration signals will include the dynamic response of each element. Each response will contribute to the overall signal and, for linear systems, they can be associated with a specific frequency. In a gear transmission there are two main sources of vibration: torsional vibrations and radial vibrations. Torsional vibrations are caused by the power supply (torque variations) and by the rotating elements. As was shown in Chapter 5, torsional vibrations are related to variations in the gearing stiffness while radial vibrations are produced by a shaft's flexibilities and the stiffness of the supports (either rolling bearings or journal bearings).

Measuring torsional vibrations requires special instrumentation, and it can only be measured without rotating the shafts. Radial vibrations can be measured rotating the shaft since the vibration transducers can be mounted on the casing.

In this example, the excitation force was applied on the input shaft and the output response was recorded at the output shaft. Both shafts were tight in order to allow only small vibrations. With an instrumented hammer, an impact torque was applied to the input shaft and the torsional accelerations were recorded with a couple of

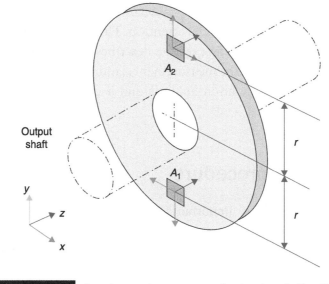

Figure 6.1 Accelerometers arrange for torsional vibrations

accelerometers mounted on the output shaft as shown in Figure 6.1.

With this arrangement, three accelerations were obtained:

$$A_r = \frac{A_{2y} - A_{1y}}{2}$$

$$A_a = \frac{A_{2z} + A_{2z}}{2}$$

$$\alpha = \frac{A_{2x} + A_{1x}}{2r} \qquad [6.1]$$

where A_r is the radial acceleration, A_a is the axial acceleration and α is the torsional acceleration. These equations were calculated simultaneously and they were synchronized with the hammer signal. In this way the sensitivity differences of each accelerometer are compensated. Sensitivity values are listed in the following table:

	Sensitivity
Accelerometer 1	$x = 10.15\,mV/g$
	$y = 10.17\,mV/g$
	$z = 10.21\,mV/g$
Accelerometer 2	$x = 9.75\,mV/g,$
	$y = 10.04\,mV/g,$
	$z = 10.23\,mV/g$
Impact hammer	$22.7\,mV/N$

Figure 6.2 shows the excitation signal. The signal damps in a very short time, therefore it is considered as an impact force or a pick force. This experiment is equivalent to the time domain analysis described in Chapter 4.

Figure 6.3 shows the output signals. Figure 6.3a corresponds to A_1y and A_2y and Figure 6.3b to α.

According to methods described in Chapter 5, the nonlinear behavior was evaluated using the phase diagram and the time–frequency maps. Since the output signal is acceleration, then phase diagram (phase plane) was constructed obtaining the angular velocity ω and rotation θ integrating the output response in the frequency domain, and then reproducing both signals in the time domain. Figure 6.4 shows the evolution of the phase diagram as a function of time:

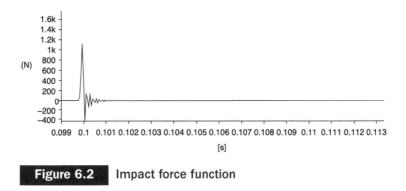

Figure 6.2 Impact force function

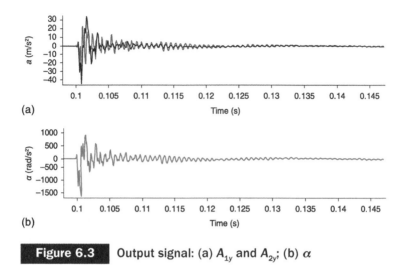

(a)

(b)

Figure 6.3 Output signal: (a) A_{1y} and A_{2y}; (b) α

Figure 6.4a corresponds to the initial time interval; Figures 6.4b and 6.4c are obtained during a transition stage, when the signal is decrementing (Figure 6.4c shows a pole outside the origin and it starts developing rough shapes at the extreme borders in the horizontal direction); Figure 6.4d corresponds to the final stage (from this graph it is clear that the system is not stable and there are two dominant poles outside the origin).

Similar conclusions were obtained from the time–frequency map. These maps were constructed with the Continuous Wavelet Transform and the mother wavelet (Chapter 3) was the Morlet function.

From the frequency spectrum, it was possible to appreciate the dynamic response in radial, axial and torsional directions. Figure 6.5 shows the frequency spectrum for the torsional vibration. This figure shows the dominant peak at 1357 Hz, and other peaks that are not synchronous.

Figure 6.6 shows the frequency spectrum in the radial direction. For this direction, the dominant peak appears at

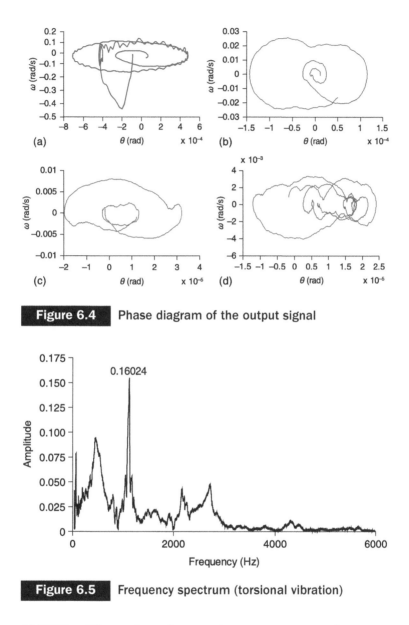

Figure 6.4 Phase diagram of the output signal

Figure 6.5 Frequency spectrum (torsional vibration)

3098 Hz. The other frequencies are not synchronous; therefore they correspond to different elements or masses. It is interesting to see that around each peak there is a wide range of frequencies.

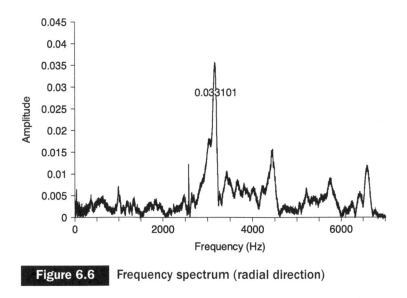

Figure 6.6 Frequency spectrum (radial direction)

Figure 6.7 shows the frequency spectrum for the radial direction. Two types of movement produce this vibration, a pure axial displacement and a shaft's bending deformation. In this case, there are three dominant peaks, one at 970 Hz, the second one at 1744 Hz and the third one at 3098 Hz. This last peak is present in all directions.

In summary, the dominant peaks have the following frequencies:

Direction	Frequency (Hz)
Torsional	583.3
	1357
	2517
	3098
Radial	3098
	4259
	5613
Axial	970
	1744
	3098

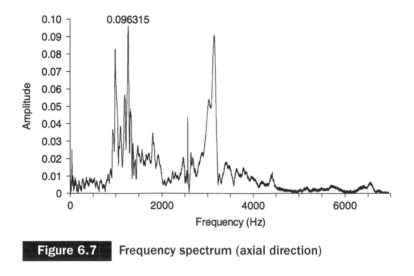

Figure 6.7 Frequency spectrum (axial direction)

It is complicated to draw a conclusion from these values. Therefore, time–frequency maps were constructed in order to analyze the behavior of each frequency response. It is important to analyze each signal around the dominant frequencies, and from variations along the time dimension it is possible to determine which frequencies contribute to the nonlinear behavior. As was detailed in previous chapters, linear responses will display continuous horizontal strips along the time dimension, whereas nonlinear responses have irregular figures along the time axis.

Figure 6.8 is the time–frequency map corresponding to the torsional vibration. There are five dominant peaks. Observing the map, it is clear that the response around 1357 Hz have a nonlinear response. From this map it is possible to determine the damping coefficients for each frequency. This time–frequency map confirms the phase diagram presented in Figure 6.4

Figure 6.9 is the time–frequency map of the radial vibration. In this case the dominant peak is at 3098 Hz and

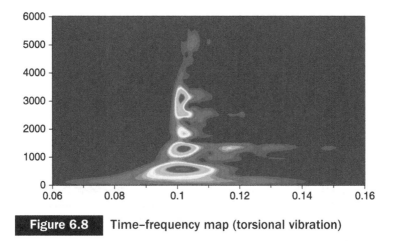

Figure 6.8 Time–frequency map (torsional vibration)

it corresponds to the output shaft's radial motion. It also shows a light nonlinear behavior.

Axial vibration (Figure 6.10) combines both motions, bending and the gear's helix angle. This map confirms that there are two nonlinear responses: torsional vibrations, due to teeth nonlinear stiffness; and radial vibrations, due to rolling bearing support.

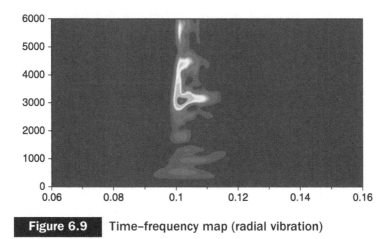

Figure 6.9 Time–frequency map (radial vibration)

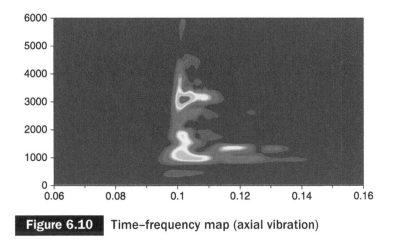

Figure 6.10 Time–frequency map (axial vibration)

These maps allow us to simulate the transmission behavior. For this purpose, equation 5.23 (Chapter 5) was solved assuming an equivalent impact force. Simulation parameters were determined from the design data and from the time–frequency maps.

6.3 Simulation analysis

Using the same model presented in Chapter 5, the equations of motion are:

$$\ddot{x}_1 + \frac{P_1}{m_1} + \frac{S_t}{m_1}[(x_1 - x_2) + (r_1\theta_1 + r_2\theta_2)] = 0$$

$$\ddot{x}_2 + \frac{P_2}{m_2} - \frac{S_t}{m_2}[(x_1 - x_2) + (r_1\theta_1 + r_2\theta_2)] = 0$$

$$\ddot{\theta}_1 + \frac{S_t r_1}{J_1}[(x_1 - x_2) + (r_1\theta_1 + r_2\theta_2)] = 0$$

$$\ddot{\theta}_2 - \frac{S_t r_2}{J_2}[(x_1 - x_2) + (r_1\theta_1 + r_2\theta_2)] = 0 \qquad [6.2]$$

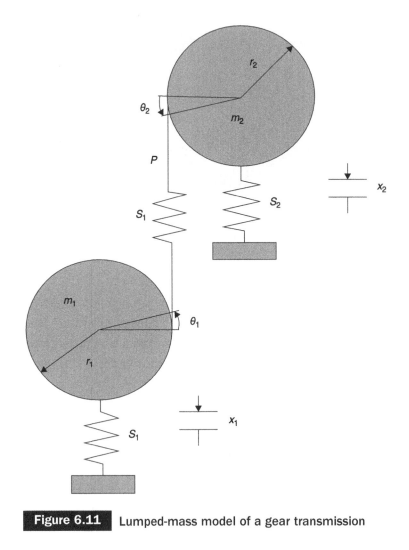

Figure 6.11 Lumped-mass model of a gear transmission

where x_1 is the radial displacement of the input shaft, x_2 is the radial displacement of the output shaft, θ_1 is the angular displacement of the input shaft and θ_2 is the angular displacement of the output shaft.

Where P_i is the roller bearing restoring force and S_t is the gear mesh stiffness (Chapter 5):

$$P_i = E\sqrt{\frac{Dd}{D-d}\left(\frac{\delta_i}{\alpha}\right)^3}$$ [6.3]

$$S_t = \begin{cases} 2Eb\tan\left(\dfrac{\alpha}{2}\right)^3 (n+1), 0 < \theta < \dfrac{2\pi}{N}(m_p - n) \\ 2Eb\tan\left(\dfrac{\alpha}{2}\right)^3 (n+1), \dfrac{2\pi}{N}(m_p - n) < 0 < \dfrac{2\pi}{N} \end{cases}$$ [6.4]

The simulation was solved using the Runge-Kutta method and the particular values are listed in the following table:

Magnitude	Value
m_1	6.5112 kg
m_2	8.311 kg
J_1	1.3454×10^3 kg-m^2
J_2	5.718410^3 kg-m^2
r_1	42.96110^3 m
r_2	117.62510^3 m
D_1	25×10^3 m
d_1	5×10^3 m
D_2	25×10^3 m
d_2	5×10^3 m
E	210000×10^6 N/m^2
m_p	4.403

With these data, we obtained the following results. For comparison purposes, only the radial displacement and torsional vibrations for the output shaft are included.

Two stages of the phase diagram are included in Figure 6.12. The phase diagram is constructed with the torsional vibration only. In the first stage, it can be noticed that two poles start to form and in the second stage they are fully formed but are outside from the origin. Although in the analysis of the

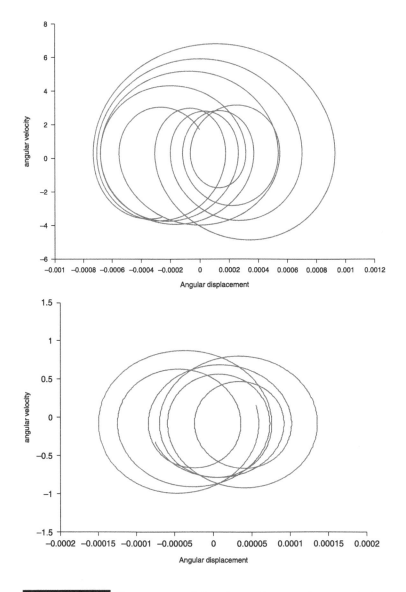

Figure 6.12 Two stages of the phase diagram (output shaft)

experimental data the shape shows more variations, it is clear that there is a nonlinear response.

The frequency spectrums are included in Figures 6.13 and 6.14. Figure 6.13 shows the frequency spectrum in the radial direction and Figure 6.14 corresponds to the torsional vibrations. Both spectrums show four dominant peaks. In the case of the radial vibration the highest peak occurs at 602 Hz, whereas in the torsional case the highest peak is at 1429 Hz. It is also important to distinguish that the peaks show a wide frequency band base, this is another indication of the nonlinear behavior.

Figure 6.15 and 6.16 show the corresponding time–frequency maps. In both cases the nonlinear response is observed. It is important to point out that, in the case of radial vibration, the dominant frequency decreases as a function of time. This frequency decrement is due to the fact that roller stiffness depends on the deformation (equation

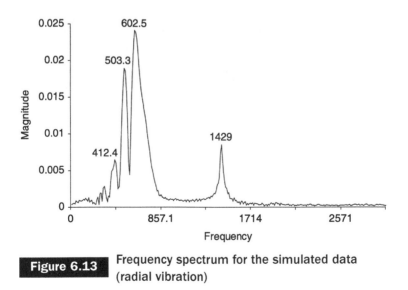

Figure 6.13 Frequency spectrum for the simulated data (radial vibration)

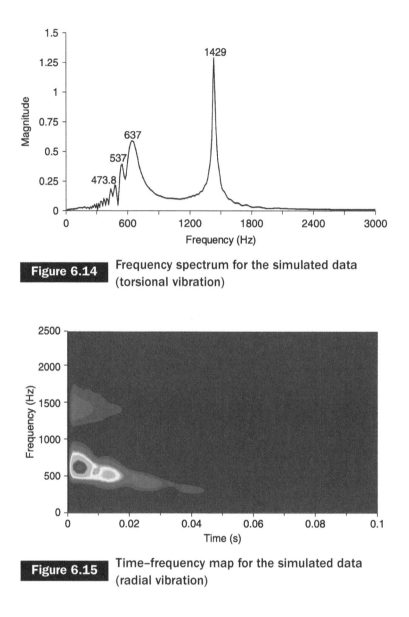

Figure 6.14 Frequency spectrum for the simulated data (torsional vibration)

Figure 6.15 Time–frequency map for the simulated data (radial vibration)

6.3). The damping effect and the nonlinear response can also be seen. The damping effect is reflected as a decrement in amplitude and the nonlinear effect is reflected as changes in amplitude as a function of time.

Figure 6.16 is the time–frequency map obtained from the torsional response. Comparing Figures 6.16 and 6.17 it is easy to find the frequency associated with the radial displacement and torsional vibrations. With this example several conclusions are drawn. The identification of nonlinear parameters requires various other analysis techniques than the well-known linear parameter identification. Additionally to the frequency spectrum, in this case we analyzed the data with the phase diagram and the time–frequency map. The frequency spectrum helps in the identification of the dominant frequencies and the range of analysis. The phase diagram gives information regarding the type of nonlinearity, for example, the presence of two poles can be associated to a stiffness function that depends on the displacement, whereas a distorted shape could be related to other nonlinearities. Finally, constructing time–frequency maps with the Continuous Wavelet Transform and the Morlet mother function allows us to identify frequency variations along the

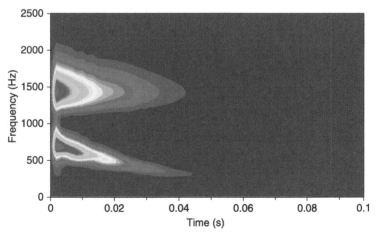

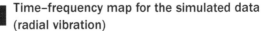

Figure 6.16 Time–frequency map for the simulated data (radial vibration)

time domain. With this information, it is possible to relate each element of the system to a corresponding degree of freedom. For nonlinear cases, the best simulation is obtained with a lumped-mass model. The type of function and values for the stiffness and damping coefficients are obtained from the three analyses.

7

Synchronization of nonlinear systems

DOI: 10.1533/9781782421665.177

Abstract: This chapter complements previous concepts because any mechanical system is a complex system. Additionally to the nonlinear analysis of mechanical systems, synchronization represents a new area of interest because there are unsolved issues regarding the interaction of individual elements in complex systems. In many applications, engineering knowledge has demonstrated limitations in the identification of roots of failure, and some of them could be related to travelling impulses among machine elements. Synchronization concepts, analysis tools and Kuramoto's parameter are described. Kuramoto's parameter is a simple and useful method for determining if different components move synchronously, and it helps in the identification of those elements that interact within a complex system. To illustrate these concepts, application examples are included for both field data and simulation models.

Key words: synchronizations, Kuramoto's parameter, complex systems.

7.1 Introduction

We have seen that mechanical systems cannot be analyzed as simple systems; they are complex systems with interactions among individual elements whose behavior is difficult to understand. Up to now, we have been able to discretize complex systems as a set of a few masses connected by elastic elements, or we have been able to model continuous systems with finite elements; but these approaches have failed to understand some of the interactions among individual elements which are not clearly defined. Therefore, it is important to broaden the analysis techniques to introduce new concepts that could help us understand the effect of those unclear defined connections.

If we represent a mechanical system as a set of interconnected elements (Figure 7.1), we can identify two types of

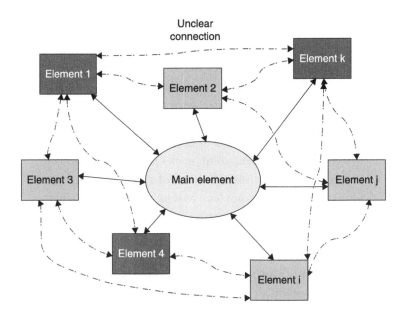

Figure 7.1   Schematic representation of the interactions within a complex system

connections: strong or dominant connections (solid line) and weak or unclear connections (dashed line). Traditionally, we model a mechanical system considering only dominant connections. But in complex systems, sometimes there are phenomena difficult to understand such as synchronization. In this chapter, nonlinear synchronization among elements is presented; this concept can be related to the effect of unclear connections in a complex systems. In previous chapters, we presented the analysis of complex systems considering only the dominant connections; adding this concept, a general view of the dynamic of complex systems is presented.

7.2 Synchronization

Synchronization is revealed as a group of elements, or systems, with a collective coherent response that can be in phase for a long period of time. For example, people filling an auditorium will start applauding at an incoherent rhythm, but a few seconds later the auditorium will have uniform sound because all the applause will synchronize. Synchronization will happen regardless of the strength of each individual beat. In this example, we can consider that each individual response is autonomous, but there is weak connection among individuals that makes them to applause in phase. Therefore, in such systems, a relatively small force can adjust phase and frequency without affecting the amplitude; this adjustment is in essence the source of synchronization. The question is what causes this force, and how this force is transmitted. Many examples exist in nature; people walking on a soft bridge, the flapping of a group of birds, cardiac pacemakers, turbine rotor blades, similar machines in a production line, etc. In this type of systems the individual elements are very similar, but they are mounted on relatively soft structures. Therefore, according to the type of structure and the connection among them, we can identify two types of synchronization: slave–master synchronization or structural synchronization. In the slave–master type, there is a dominant element setting the rhythm and the remaining elements adjust to it. In the structural synchronization, the phenomenon depends upon the structural stiffness that interconnects all the individual elements.

The relevance of synchronization has been pointed out by many researchers, although it has not always been fully understood. The main interest is establishing the mechanism responsible of connecting and transmitting impulses among individual elements. To achieve a coherent activity, or movement, each element must have an oscillatory response linked at least to another element. The rhythmical activity may be due to external stimuli transmitted along the element's linkage. There is a cyclic mechanism in the stimuli emission of bidirectional nature: the stimuli is emitted at certain periods, and will be received by an element. This element will retransmit it at a different period through the linkage until all the elements move at a coherent rhythm. Self-sustained synchronization in mechanical systems has been of growing interest, but there is limited literature in comparison with other synchronization problems. A classic case is that of Huygens' pendulums: he observed, in 1665, that the pendulums of two similar clocks mounted on a common support will tend to synchronize due to structural interaction.

There are different points of view regarding how to determine synchronization. One of them is evaluating if the relative displacement of every element asymptotically tends to zero. Another point of view establishes that synchronization occurs when the phase angle remains constant.

The most successful attempt was due to Kuramoto, who analyzed a model of phase oscillators running at arbitrary intrinsic frequencies, and coupled through the sine of their phase differences. Kuramoto's approach is a simple mathematical model that sets a feasible parameter for evaluating synchronization.

7.3 Kuramoto's model

The Kuramoto model corresponds to the simplest possible case of equally weighted, purely sinusoidal coupling. It is based on a set of N coupled oscillators, $\theta_i(t)$, having natural frequencies $\omega_i(t)$ with a constant amplitude (Figure 7.2).

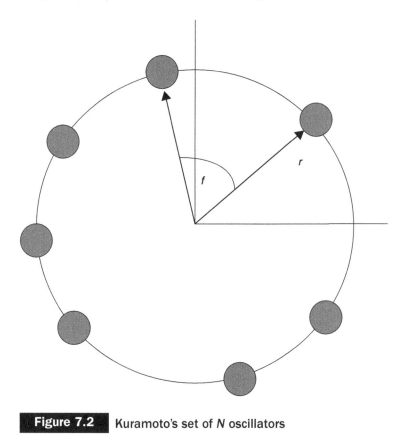

Figure 7.2 Kuramoto's set of N oscillators

Following the idea of an impulse being transmitted between oscillators, then there is a dynamic relationship of the form:

$$\frac{d\theta_i}{dt} = \omega_i + \sum_i^N K_{ij} \sin(\theta_i - \theta_j) \qquad [7.1]$$

If we consider that ω_i has a probabilistic distribution, then we can transform equation 7.1 into an equivalent system of phase oscillators whose natural frequencies have zero mean. When the coupling is sufficiently weak, the oscillators move incoherently, but there is a certain threshold coupling at which collective synchronization emerges spontaneously. Many different models for the coupling matrix K_{ij} have been considered, such as nearest-neighbor coupling, hierarchical coupling, random long-range coupling, or even state dependent interactions.

Defining the average angle between each relative position

$$re^{i(\varphi-\theta_i)} = \frac{1}{N} \sum_{j=1}^N e^{i(\theta_j - \theta_i)} \qquad [7.2]$$

From the definition of synchronization, at a certain stage the relative phase angle will tend to a zero mean or to a constant phase φ. Substituting the imaginary part of equation 7.2 into equation 7.1

$$\frac{d\theta_i}{dt} = \omega_i + rK_{ij} \sin(\phi - \theta_j) \qquad [7.3]$$

This result shows that the order parameter r and phase φ summarizes the effect that all the oscillators have on any particular oscillator. There is a locked frequency Ω at which all the oscillators will have the same phase rate. Therefore, at a steady state condition $\dfrac{d\theta_i}{dt} = 0$

$$\omega_i = \Omega + rK_{ij}\sin(\theta_j) \qquad [7.4]$$

In this way, the oscillators whose frequencies fall within the following range will have a locked phase.

Kuramoto suggested a critical parameter Kc derived from a continuous variable, assuming that it follows a probabilistic density function $g(\omega)$ such that

$$r = \int_{-\pi/2}^{\pi/2} e^{i\theta} g(\omega) \frac{d\omega}{d\theta} d\theta \qquad [7.5]$$

From the steady state solution (equation 7.4)

$$\frac{d\omega}{d\theta} = rK_{ij} \cos(\theta) \qquad [7.6]$$

Expanding the Euler's number, and assuming that the density function is symmetric around Ω:

$$K_{ij} = \frac{1}{\int_{-\pi/2}^{\pi/2} g(\omega) \cos^2(\theta) d\theta} \qquad [7.7]$$

The oscillators with a frequency close to the locking frequency Ω will have a critical value of

$$K_c = \frac{2}{\pi g(\Omega)} \qquad [7.8]$$

Assuming the $g(\Omega)$ follows a Lorentzian distribution, it can be demonstrated that

$$r = \sqrt{1 - \frac{K}{K_c}} \qquad [7.9]$$

In general, for a set of N interconnected elements with similar natural frequencies, the Kuramoto parameter can be estimated as:

$$r(t) = \frac{1}{N} \left[\sum_{j=1}^{N} e^{i\varphi(t)_j} \right] \qquad [7.10]$$

where φ_j is the instantaneous phase angle between each oscillator.

If we measure the dynamic response of each element and we store the data in a vector $x(t)$, the instantaneous phase angle can be determined using the Hilbert transform, defining as:

$$y(t) = \int_{-\infty}^{\infty} \frac{x(t)}{\pi(t-u)} du \qquad [7.11]$$

Combining the original data with its Hilbert transform into a complex number:

$$z(t) = x(t) + iy(t) \qquad [7.12]$$

Then, the instantaneous phase can be determined as:

$$\tan(\varphi(t)) = \frac{y(t)}{x(t)} \qquad [7.13]$$

Two examples are presented next. The first example is the measurement of two torsional accelerometers mounted on the opposite sides of a motor. The relative change in phase is proportional to a change in torque; thus using Kuramoto's parameter it is possible to estimate any change in torque. The second example shows the synchronization phenomenon of a set of 22 blades excited with an air flow.

7.3.1 Phase difference in an electric motor shaft

Torsion variations in a shaft can be determined by measuring the phase angle between the two ends of the shaft. Based on this principle, two angular accelerometers where mounted at each end of an electric motor. In this way, instantaneous torque variations can be determined identifying phase variation between the two signals. There is an angular shift between the two accelerometers as noted in Figure 7.3. In this figure the actual angular acceleration is plotted and shows a sinusoidal function with high frequency variations

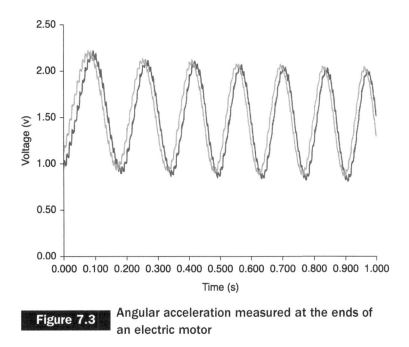

Figure 7.3 Angular acceleration measured at the ends of an electric motor

due to measurement uncertainties. These data are a good approximation to an ideal harmonic function, and they illustrate the effect of those measurement uncertainties on Kuramoto's parameter.

Kuramoto's parameter determines the amount of synchronization in a system with several oscillators. The instantaneous phase was calculated with the Hilbert's transform (equation 7.11) and the phase difference is plotted in Figure 7.4. In this case, there are only two signals, therefore, $N = 2$ and Kuramoto's parameter varies with time as displayed in Figure 7.5. It is clear that all the values are above 0.985, which means that both signals are in phase. Kuramoto's parameter shows periodic variations, these variations are the result of measuring deviations from a pure harmonic signal.

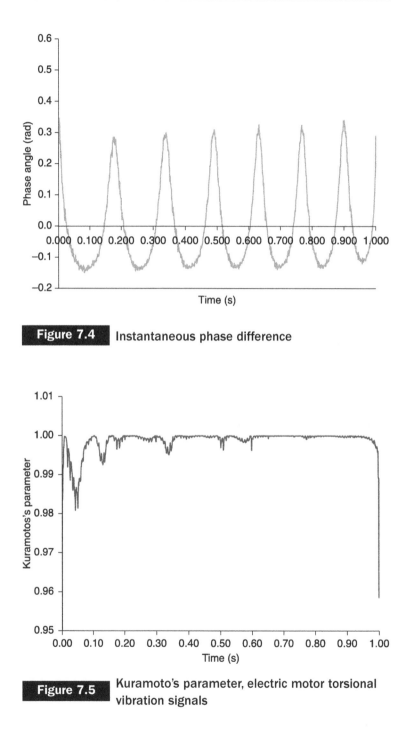

7.3.2 Turbine blades under an airflow stream

In the next example, 22 accelerometers were mounted on the tip of 22 rectangular strips representing a turbine rotor. Each strip was identical and was mounted on a fixed shaft (the shaft was stationary without rotation Figure 7.6) and was excited by an airstream. The individual natural frequency was around 121 Hz. Turbine blades in a single rotor may synchronize either because of the influence of the fluid or by structural interaction. In this case, the effect of the wind was considered only as an external excitation and the interaction among the blades is assumed to occur through the structure.

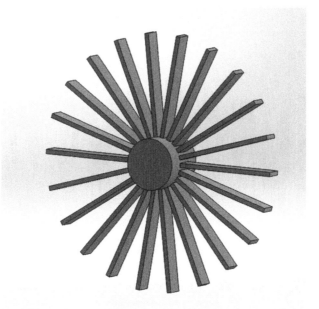

 Sketch of 22 blades, the wind flows in the radial direction

Individual accelerations were recorded at different time intervals, and Kuramoto's parameter was calculated for data vectors of two seconds. The instantaneous phase was calculated with the Hilbert's transform (equation 7.11) and $N = 22$. At every interval, Kuramoto's parameter behavior was similar to the one presented in Figure 7.7. In this figure we use a large scale for the amplitude; it is evident that the amplitude varies frequently, but the amount of variation is very small. If we calculate the frequency spectrum with the Fourier Transform, Kuramoto's parameter shows a dominant frequency around twice the blades' natural frequency in Figure 7.8. Since r is above 0.997 we can assume that there is a strong connection. This example illustrates a case of structural synchronization.

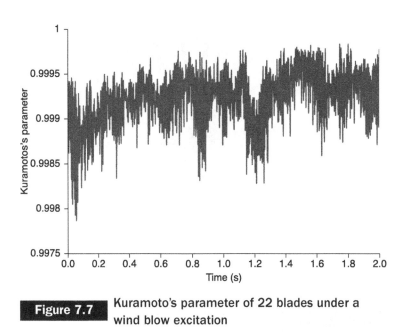

Figure 7.7 Kuramoto's parameter of 22 blades under a wind blow excitation

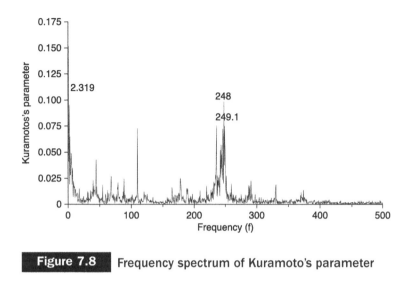

Figure 7.8 Frequency spectrum of Kuramoto's parameter

7.4 Synchronization of nonlinear pendulums

Even if a mechanical system has elements of a different nature, it may also show synchronization phenomena. As we have seen in previous chapters, nonlinear systems can be represented as an ensemble of nonlinear oscillators moving around attracting limit cycles; these limit cycles can be considered as phase limitations. These oscillators are coupled through weak or strong elements; for example, in a gear box, the gear pair can be considered as a nonlinear oscillator and the bearings as another oscillator, in this case the casing is the coupling element. The stiffness of the casing will determine if the coupling is weak or strong. A weak coupling enables the oscillator to move away from the limit cycle, whereas a strong coupling synchronizes all the oscillators around the limit cycle. It is not easy to propose specific models; one approach is the identification of pulse coupling, but the outcome is complicated mathematical models, and this description is more complicated for complex systems with a large number of nonlinear elements. The following example helps understand the interaction of nonlinear elements coupled through weak and strong connectors.

The example shows synchronization of two nonlinear oscillators: a Duffing oscillator coupled to a Van der Pol oscillator through a linear spring. It is assumed that a Van der Pol oscillator behaves similarly to a roller bearing and the Duffing oscillator resembles the dynamics of a pair of gears. The casing is represented as a linear spring. This simple model represents a complex system with multiple degrees of freedom, it focuses on the response of the nonlinear elements that dominate the dynamic response of the entire system. Figure 7.9 shows a simplified sketch of the system.

In order to study synchronization between two nonlinear oscillators a simple model is proposed (Figure 7.9). The model consists of two nonlinear pendulums of equal mass m and equal length l interconnected by a linear spring k. One

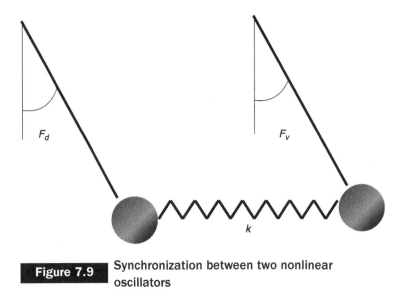

Figure 7.9 Synchronization between two nonlinear oscillators

of the oscillators is assumed to behave as a Duffing's type oscillator and the other as a Van der Pol's oscillator.

For the Duffing's term:

$$\frac{d^2\varphi_d}{dt^2} + a\frac{d\varphi_d}{dt} + \omega_n^2\varphi_d + \varepsilon\varphi_d^3 + \frac{k}{m}(\varphi_d - \varphi_v) = 0 \qquad [7.14]$$

And, for the Van der Pol's term, the dynamic equation is:

$$\frac{d^2\varphi_v}{dt^2} + \mu(1-\varphi_v^2)\frac{d\varphi_v}{dt} + \omega_n^2\varphi_v + \frac{k}{m}(\varphi_v - \varphi_d) = 0 \qquad [7.15]$$

7.4.1 Strong connection

Since we want to analyze how the connector affects synchronization, two solutions were found. The first solution considers a strong connection. For this case, k/m is larger than ω_n^2. Physically, it could represent a very stiff housing. Solving for φ_v and φ_d and applying equation 7.10 it is possible to compute Kuramoto's parameter as a function of time (the calculation is similar to previous examples). Figure 7.10 shows the behavior of this parameter. It is interesting to notice that there is a transition period where oscillators synchronize, and then they move independently. The effect of this period of synchronization can cause an additional dynamic loading on both elements and it can occur every time that there is a change of conditions, such as change on torque.

7.4.2 Weak connection

The weak connection is simulated considering that k/m is smaller than ω_n^2. Physically, it could represent a soft housing. Solving for φ_v and φ_d and applying equation 7.10 it is possible to compute Kuramoto's parameter as function of time (the calculation is similar to previous examples). Figure 7.11

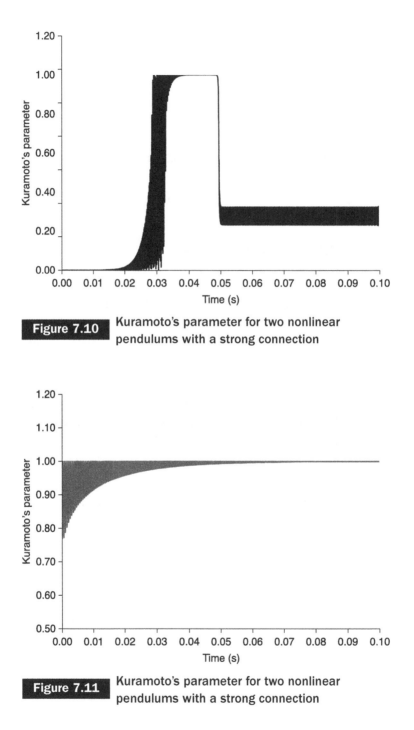

Figure 7.10 Kuramoto's parameter for two nonlinear pendulums with a strong connection

Figure 7.11 Kuramoto's parameter for two nonlinear pendulums with a strong connection

shows the behavior of this parameter. The behavior is completely different from the strong connection case. It takes only half of the time to reach full synchronization (r close to 1). In this case, full synchronization is reach asymptotically and both oscillators move in phase. In a real system, such as a gear box, full synchronization will mean that small impulses are transmitted through the casing. If the casing is soft, such impulses will create travelling waves that will increase the dynamic loading. Comparing the results shown in Figures 7.10 and 7.11, a weak connection will allow synchronization during larger periods of time. This condition will increase the dynamic loading due to the travelling waves (we can name them as synchronization impulses) and these impulses will reduce the effective life of all the elements of the system. Accounting for these additional dynamic loads is an area of future research.

Bibliography

Abbes, M., Trigui, M., Chaari, F., Fakhfakh, T. and Haddar, M. (2010) Dynamic behavior modelling of a flexible gear system by the elastic foundation theory in presence of defects, *European Journal of Mechanics. A/Solids*, Vol. 29, pp. 887–896.

Acebron, J., Bonilla, L., Pérez, C., Ritort, F. and Spigler, R. (2005) The Kuramoto model: A simple paradigm for synchronization phenomena, *Review of Modern Physics*, Vol. 77, No. 1, pp. 137–185.

Anderson, J. and Ferri, A. (1990) Behavior of a single-degree-of-freedom-system with a generalized friction law, *Journal of Sound and Vibration*, Vol. 140(2), pp. 287–304.

Andreaus, U. and Baragatti, P. (2011) Cracked beam identification by numerically analyzing the nonlinear behaviour of the harmonically forced response, *Journal of Sound and Vibration*, Vol. 330, pp. 721–742.

Barron, M. and Sen, M. (2009) Synchronization of four coupled van der Pol oscillators, *Nonlinear Dynamics*, Vol. 56, pp. 357–367.

Bartelmus, W., Chaari, F., Zimroz, R. and Haddar, M. (2010) Modelling of gearbox dynamics under time-varying non-stationary load for distributed fault detection and diagnosis, *European Journal of Mechanics. A/Solids*, Vol. 29, pp. 637–646.

Bracewell, R. (1999) *The Fourier Transform and its Applications*, 3rd edn. McGraw-Hill, New York.

Bytrus, M. and Zennan, V. (2011) On the modeling and vibration of gear drives influenced by nonlinear coupling, *Mechanisms and Machine Theory*, Vol. 46, pp. 375–397.

Chang, C. (2010) Strong nonlinearity analysis for gear-bearing system under nonlinear suspension – bifurcation and chaos, *Nonlinear Analysis: Real World Applications*, Vol. 11, pp. 1760–1774.

Changenet, C., Oviedo-Marlot, X. and Velex, P. (2006) Power loss predictions in geared transmissions using thermal networks-applications to a six-speed manual gearbox, *Journal of Mechanical Design*, Vol. 128, No. 3, pp. 618–625.

Chui, C. (1992) *An Introduction to Wavelets*. Academic, New York.

Cohen, L. (1989) Time-frequency distributions – a review. *Proceedings of the IEEE*, Vol. 77, No. 7, pp. 941–981.

Cooley, J. and Tukey, J. (1965) An algorithm for the machine calculation of complex Fourier series. *Math Comput*, Vol. 19, pp. 297–301.

Daubechies, I. (1988) Orthonormal bases of compactly supported wavelets. *C ommunications on Pure and Applied Mathematics*, Vol. 4, pp. 909–996.

Daubechies, I. (1992) *Ten Lectures on Wavelets*. SIAM, Philadelphia, PA

De Santiago, O. and San Andres, L. (2007) Experimental Identification of bearing dynamic force coefficients in a flexible rotor – further developments, *Tribology Transactions*, Vol. 50(1), pp. 114–126.

De Santiago, O. and San Andres, L. (2007) Field methods for identification of bearing support parameters–Part I: Identification from transient rotor dynamic response due to impacts, *Journal of Engineering for Gas Turbines and Power*, Vol. 129, pp. 205–212.

DeVore, R., Jawerth, B. and Lucier, B. (1992) Image compression through wavelet transform coding. *IEEE Transactions on Information Theory*, Vol. 38, No. 2, pp. 719–746.

Diaz, S. and San Andres, L. (1999) A method for identification of bearing force coefficients and its application to a squeeze film damper with a bubbly lubricant, *Tribology Transactions*, Vol. 42 (4), pp. 739–746.

Faggioni, M., Samani, F., Bertacchi, G. and Pellicano, F. (2011) Dynamic optimization of spur gears, *Mechanisms and Machine Theory*, Vol. 46, pp. 544–557.

Feldman, M. (2011) Hilbert transform in vibration analysis, *Mechanical Systems and Signal Processing*, Vol. 25, pp. 735–802.

Fritzen, C. P. (1985) Identification of mass, damping, and stiffness matrices of mechanical systems, *ASME Paper* 85-DET-91.

Gabor, D. (1946) Theory of communication, *Journal of the IEEE*, Vol. 93, No. 3, pp. 429–457.

Gao, R. and Yan, R. (2011) *Wavelets, Theory and Applications for Manufacturing*, Springer, New York.

Ghafari, S., Abdel-Rahman, E., Golnaraghi, F. and Ismail, F. (2010) Vibrations of balanced fault-free ball bearings, *Journal of Sound and Vibration*, Vol. 329, pp. 1332–1347.

Giagopulos, D., Salpistis, C. and Natsiavas, S. (2006) Effect of non-linearities in the identification and fault detection of gear-pair systems, *International Journal of Non-linear Mechanics*, Vol. 41, pp. 213–230.

Goodwin, M. J. (1991) Experimental techniques for bearing impedance measurement, *Journal of Engineering for Industry*, Vol. 113, pp. 335–342.

Gopalakrishnan, S. and Mitra, M. (2010) *Wavelet Methods for Dynamical Problems*, Taylor and Francis, Boca Raton, FI.

Greenwood, D. (1977) *Classical Dynamics*, Dover Publications, New York.

Grossmann, A. and Morlet, J. (1984) Decomposition of hardy functions into square integrable wavelets of constant shape. *SIAM Journal of Mathematics Analysis*, Vol. 15, No. 4, pp. 723–736.

Grossmann, A., Morlet, J. and Paul, T. (1985) Transforms associated to square integrable group representations. I. General results. *Journal of Mathematical Physics*, Vol. 26, pp. 2473–2479.

Grossmann, A., Morlet, J. and Paul, T. (1986) Transforms associated to square integrable group representations. II: Examples. *Annales de L'Institue Henri Poincare*, Vol. 45, No. 3, pp. 293–309.

Han, B., Cho, M., Kim, C., Lim, C. and Kim, J. (2009) Prediction of vibrating forces on meshing gears for a gear rattle using a new multi-body dynamic model, *International Journal of Automation Technology*, Vol. 10, No. 4, pp. 469–474.

He, W., Jiang, Z. and Qin, Q. (2010) A joint adaptive wavelet filter and morphological signal processing method for weak mechanical impulse extraction, *Journal of Mechanical Science and Technology*, Vol. 24, No. 8, pp. 1709–1716.

Herivel, J. (1975) *Joseph Fourier. The man and the physicist.* Clarendon Press, Oxford.

Jaffard, S., Meyer, Y. and Ryan, R. (2001) *Wavelets: tools for science & technology.* Society for Industrial Mathematics, Philadelphia, PA

Jauregui, J. and Gonzalez, O. (2009) *Mechanical Vibrations of Discontinuous Systems* (1st edition), Nova Science Publishers, ISBN: 978–1-60876–126-5, New York.

Jauregui, J. and Gonzalez, O. (2009) Non-linear vibrations of slender elements, in: *Mechanical Vibrations Measurements, Effects and Control*, Sapri, R., pp. 557–588, Nova Science Publishers, New York.

Jauregui, J. (2011) Phase diagram analysis for predicting nonlinearities and transient responses, in: N. Baddour (ed.), *Recent Advances in Vibrations Analysis*, Intech, pp. 1–20.

Jauregui, J. (2013) The effect of nonlinear traveling waves on rotating machinery, *Mechanical Systems and Signal Processing*, Vol. 39, pp. 129–142.

Junsheng, C., Dejie, Y. and Yu, Y. (2007) The application of energy operator demodulation approach based on EMD in machinery fault diagnosis, *Mechanical Systems and Signal Processing*, Vol. 21, pp. 668–677.

Karagiannis, K. and Pfeiffer, F. (1991), Theoretical and experimental investigations of gearbox, *Nonlinear Dynamics*, Vol. 2, pp. 367–387.

Karpenko, E., Wiercigroch, M., Pavlovskaia, E. and Neilson, R. (2006) Experimental verification of Jeffcott rotor model with preloaded snubber ring, *Journal of Sound and Vibration*, Vol. 298, pp. 907–917.

Kerschen, G., Worden, K., Vakakis, A. and Golinvala, J. (2006) Past, present and future of nonlinear system identification in structural dynamics, *Mechanical Systems and Signal Processing*, Vol. 20, pp. 505–592.

Korkmaz, S. (2011) A review of active structural control: challenges for engineering informatics, *Computers and Structures*, Vol. 89, pp. 2113–2132.

Korner, T. (1988) *Fourier Analysis*. Cambridge University Press, London.

Li, C. and Liang, M. (2011) Time–frequency signal analysis for gearbox fault diagnosis using a generalized synchro squeezing transform, *Mechanical Systems and Signal Processing*, Vol. 26, pp. 205–217.

Li, C. and Qu, L. (2007) Applications of chaotic oscillator in machinery fault diagnosis, *Mechanical Systems and Signal Processing*, Vol. 21, pp. 257–269.

Liu, B. (2005) Selection of wavelet packet basis for rotating machinery fault diagnosis, *Journal of Sound and Vibration,* Vol. 284, pp. 567–582.

Luo, A. and O'Connor, D. (2011) Periodic and chaotic motions in a gear-pair transmission system with impacts, in: J. A. T. Machado et al. (eds), *Nonlinear Science and Complexity*, pp. 13–24.

Machado, L., Lagoudas, D. and Savi, M. (2009) Lyapunov exponents estimation for hysteretic systems, *International Journal of Solids and Structures,* Vol. 46, pp. 1269–1286.

Mackenzie, D. (2001) *Wavelets: Seeing the Forest and the Trees.* National Academy of Sciences, Washington, DC.

Mallat, S. (1989) A theory of multiresolution signal decomposition: the wavelet representation. *IEEE Trans Pattern Anal Mach Intell,* Vol. 11, No. 7, pp. 674–693.

Mallat, S. (1989) Multiresolution approximations and wavelet orthonormal bases of L2(R). *Trans Am Math Soc,* Vol. 315, pp. 69–87.

Mallat, S. (1998) *A Wavelet Tour of Signal Processing.* Academic, San Diego, CA.

Masri, S. (1994). A hybrid parametric/nonparametric approach for the identification of nonlinear systems, *Probabilistic Engineering Mechanics,* Vol. 9, pp. 47–57.

Mazzillia, C., Sanches, C., Netoa, B., Wiercigrochb, M. and Keber, M. (2008) Non-linear modal analysis for beams subjected to axial loads: Analytical and finite-element solutions, *International Journal of Non-Linear Mechanics,* Vol. 43, pp. 551–561.

Mevela, B. and Guyaderb, J. (2008) Experiments on routes to chaos in ball bearings, *Journal of Sound and Vibration,* Vol. 318, pp. 549–564.

Meyer, Y. (1989) Orthonormal wavelets. In: J. M., Combers A. Grossmann and P. Tachamitchian (eds), *Wavelets,*

Time–Frequency Methods and Phase Space, Springer-Verlag, Berlin.

Meyer, Y. (1993) *Wavelets, Algorithms and Applications.* SIAM, Philadelphia, PA.

Miao, Q., Wang, D. and Pecht, M. (2010) A probabilistic description scheme for rotating machinery health evaluation, *Journal of Mechanical Science and Technology*, Vol. 24, pp. 2421–2430.

Modarres, Y., Chasparis, F., Triantafyllou, M., Tognarelli, M. and Beynet, P. (2011) Chaotic response is a generic feature of vortex-induced vibrations of flexible risers, *Journal of Sound and Vibrations*, Vol. 330, No. 11, pp. 2565–2579.

Nagarajaiah, S. and Basu, B. (2009) Output only modal identification and structural damage detection using time–frequency and wavelet techniques, *Earthquake Engineering and Engineering Vibrations*, Vol. 8, pp. 583–605.

Newland, D. (1993) Harmonic wavelet analysis, *Proc Royal Society of London, A Math Phys Sci*, Vol. 443, pp. 203–225.

Nichols, J. (2003) Structural health monitoring of offshore structures using ambient excitation, *Applied Ocean Research*, Vol. 25, pp. 101–114.

Nordmann, R. and Schollhorn, K. (1980) Identification of stiffness and damping coefficients of journal bearings by means of the impact method, *Proc. Vibrations in Rotating Machinery: 2nd International Conference*, Inst. Mech. Eng., Cambridge, UK, pp. 231–238.

Oden, J. and Martins, J. (1985) Models and computational methods for dynamic friction phenomena, *Computer Methods in Applied Mechanics and Engineering*, Vol. 52, pp. 527–634.

Ognjanovic, M. and Benur, M. (2011) Experimental research for robust design of power transmission components, *Meccanica*, Vol. 46, pp. 699–710.

Oppenheim, A., Schafer, R. and Buck, J. (1999) *Discrete Time Signal Processing*. Prentice Hall PTR, Englewood Cliffs, NJ.

Ott, E. (2002) *Chaos in Dynamical Systems*, 2nd edn. Cambridge University Press, U.K.

Ozguven, H. and Houser, D. (1988) Mathematical models used in gear dynamics—a review, *Journal of Sound and Vibration*, Vol. 121, No. 3, pp. 383–411.

Pai, F. (2007) Nonlinear vibration characterization by signal decomposition, *Journal of Sound and Vibration*, Vol. 307, pp. 527–544.

Park, S., Sim, H., Lee, H. and Oh, J. (2008) Application of non-stationary signal characteristics using wavelet packet transformation, *Journal of Mechanical Science and Technology*, Vol. 22, pp. 2122–2133.

Qian, S. (2002) *Time–Frequency and Wavelet Transforms*. Prentice Hall PTR, Upper Saddle River, NJ

Rafiee, J., Rafiee, M. and Tse, P. (2010) Application of mother wavelet functions for automatic gear and bearing fault diagnosis, *Expert Systems and Applications*, Vol. 37, pp. 4568–4579.

Rice, H. and Fitzpatrick, J. (1991) Procedure for the identification of linear and non-linear multi-degree-of-freedom systems, *Journal of Sound and Vibration*, Vol. 149, No. 3, pp. 397–411.

Ricker, N. (1953) The form and laws of propagation of seismic wavelets, *Geophysics*, Vol. 18, pp. 10–40.

Rioul, O. and Vetterli, M. (1991) Wavelets and signal processing, *IEEE Signal Process Mag*, Vol. 8, No. 4, pp. 14–38.

Rouvas, C. and Childs, D. (1993) A parameter identification method for the rotordynamic coefficients of a high speed Reynolds Number hydrostatic bearing, *Journal of Vibration and Acoustics*, Vol. 115, pp. 264–270.

Rubio, E. and Jauregui, J. (2011) Time–frequency analysis for rotor-rubbing diagnosis, *Advances in Vibration Analysis Research*, Ebrahimi, F., InTech Publishers.

Rüdinger, F. and Krenk, S. (0000) Non-parametric system identification from non-linear stochastic response, *Probabilistic Engineering Mechanics*, Vol. 16, pp. 233–243.

San Andrés, L. and Delgado, A. (2007) Identification of force coefficients in a squeeze film damper with a mechanical end sea – centered circular orbit tests, *Journal of Tribology*, Vol. 129, No. 3, pp. 660–668.

Schuëller, G. (1997) A state-of-the-art report on computational stochastic mechanics, *probabilistic engineering mechanics*, Vol. 12, No. 4, pp. 197–321.

Setter, E. and Bucher, I. (2011) Flexural vibration patterning using an array of actuators, *Journal of Sound and Vibration*, Vol. 330, pp. 1121–1140.

Strogatz, S. (2000) From Kuramoto to Crawford: exploring the onset of synchronization in populations of coupled oscillators, *Physica D*, Vol. 143, pp. 1–20.

Stromberg, J. (1983) A modified Franklin system and higher-order spline systems on Rn as unconditional bases for Hardy space. *Proceedings of Conference on Harmonic Analysis in Honor of Antoni Zygmund*, Vol. 2, pp. 475–494.

Su, Z., Zhang, Y., Jia, M., Xu, F., and Hu, J. (2011) Gear fault identification and classification of singular value decomposition based on Hilbert–Huang transform, *Journal of Mechanical Science and Technology*, Vol. 25, No. 2, pp. 267–272.

Tiwari, R., Lees, A. and Friswell, M. (2004) Identification of dynamic bearing parameters: a review, *Shock and Vibration Digest*, Vol. 36, pp. 99–124.

Vela, L., Jauregui, J., Rodriguez, E. and Alvarez, J. (2010)

Using detrended fluctuation analysis to monitor chattering in cutter tool machines, *International Journal of Machine Tools & Manufacture*, Vol. 50, pp. 651–657.

Wang, C., Fang, Z. and Jia, H. (2010) Investigation of a design modification for double helical gears reducing vibration and noise, *Journal of Marine Science and Applications*, Vol. 9, pp. 81–86.

Wang, G., Li, Y. and Luo, Z. (2009) Fault classification of rolling bearing based on reconstructed phase space and Gaussian mixture model, *Journal of Sound and Vibration*, Vol. 323, pp. 1077–1089.

Wang, Z., Akhtar, I., Borggaard, J. and Iliescu, T. (2011) Two-level discretizations of nonlinear closure models for proper orthogonal decomposition, *Journal of Computational Physics*, Vol. 230, pp. 126–146.

Wiercigroch, M. and Pavlovskaia, E. (2008) Non-linear dynamics of engineering systems, *International Journal of Non-Linear Mechanics*, Vol. 43, pp. 459–461.

Yang, J., Zhang, Y. and Zhu, Y. (2007) Intelligent fault diagnosis of rolling element bearing based on SVMs and fractal dimension, *Mechanical Systems and Signal Processing*, Vol. 21, pp. 2012–2024.

Yu. J., Bently, D., Goldman, P., Dayton, K. and Van Slyke, B. (2002) Rolling element bearing defect detection and diagnostics using displacement transducers, *Journal of Engineering for Gas Turbines and Power*, Vol. 124, pp. 517–524.

Index

Lightning Source UK Ltd.
Milton Keynes UK
UKOW06n1910260315

248582UK00002B/29/P